The Prehistory & History of Ceramic Kilns

Other Ceramics and Civilization volumes:

Volume VI: *The Social and Cultural Contexts of New Ceramic Technologies*
edited by W.D. Kingery
© 1993, ISBN 0-944904-61-0

Volume V: *Changing Roles of Ceramics in Society: 26 000 B.C. to Present*
edited by W.D. Kingery
© 1990, ISBN 0-944904-11-4

Volume IV: *Cross Craft and Cultural Interactions*
edited by W.D. Kingery
© 1989, ISBN 0-916094-48-0

Volume II: *Technology and Style*
edited by W.D. Kingery
© 1986, ISBN 0-916094-76-6

CERAMICS AND CIVILIZATION

The Prehistory & History of Ceramic Kilns

Volume
VII

Editor: Prudence M. Rice
Series Editor: W.D. Kingery

Proceedings of the Prehistory and History of Ceramic Kilns, held at the 98th Annual Meeting of the American Ceramic Society in Indianapolis, Indiana, April 14–17, 1996.

For information on ordering titles published by The American Ceramic Society, or to request a publications catalog, please call 614-794-5890, or visit www.ceramics.org.

2 3 4 04 03 02

ISSN 1042-1122

ISBN 978-1-57498-026-4

Contents

Foreword

This seventh volume of the *Ceramics and Civilization* series continues the efforts of The American Ceramic Society to situate and understand its activities and programs in the broad context of culture, history, and society. Ceramic sherds make up the largest fraction of material objects recovered by archaeologists from prehistoric and historic sites. As a result there is great interest in the inferences that can be developed about design, production, distribution, use, and performance from the objects themselves.

An essential component of the very definition of ceramics is the heat treatment resulting in permanent shapes and useful properties. Thus, every site where ceramics were produced contains the remains of ceramic kilns, some simple, some complex. Many archaeologists believe that too little generally available information on kiln types, construction, and use has led to their underrepresentation in excavation reports. A principal purpose of the present meeting has been to bring together ceramic engineers, ethnographers, ceramic historians, and archaeologists familiar with the normal configurations of ceramic kilns that have been developed in a wide variety of epochs, geographies, and cultures.

The simplest ceramic pots are made with a fast, low-temperature firing accomplished in a variety of configurations with no permanent structure. For larger production and longer firing permanent kilns were constructed in many different designs appropriate for particular fuels, products, raw materials, processes, environments, cultures, and social organizations. For ceramic production at all times and all places, the kiln is an integral component.

We are fortunate that the symposium was mostly organized and this volume edited by Professor Prudence M. Rice, an eminent scholar in the field of archaeological ceramics and Chair of the Department of Anthropology at Southern Illinois University. I extend to her my thanks as Series Editor and as Chairperson of the American Ceramic Society History Committee.

W. David Kingery

Preface

For many years now, The American Ceramic Society has been a wonderful venue for the presentation of stimulating symposia integrating a variety of approaches to the history of ceramics. Most of these have been organized by W. David Kingery, who has chaired the Division of Ceramic History for some years now, and is founding editor of the *Ceramics and Civilization* publication series of the Society, under whose auspices the present volume is being published. First and foremost, I wish to thank David Kingery for his encouragement and support in the organization of the kiln symposium and its publication.

I am extremely grateful to The American Ceramic Society headquarters staff, particularly Bill Douglas, who graciously accommodated the needs of a variety of nonmembers for this symposium, and to Sarah Godby, who politely tolerated my egregious violations of publication deadlines.

At the same time, I am especially indebted to the participants and attendees of this symposium, who were uniformly enthusiastic about participating in a broader exchange of information on the prehistory and history of kilns. All participants were generous with their data and incredibly receptive to the wide variety of perspectives and approaches to their study. I greatly hope that this symposium will be followed by others, in a variety of professional contexts, to continue this fruitful exploration.

I dedicate this volume to Frederick R. Matson, Professor Emeritus at The Pennsylvania State University. Fred was my advisor when I was a graduate student at Penn State, and he insisted upon an interdisciplinary and multicultural approach to pottery, incorporating readings and data from around the world, from prehistoric to modern contexts. Fred was generous with his time and his contacts, and when I as an "outsider" archaeologist attending American Ceramic Society meetings, he graciously escorted me through the meetings and introduced me to his colleagues. Fred was a wonderful mentor and several generations of ceramics students are deeply in his debt.

Prudence M. Rice

INTRODUCTION AND OVERVIEW

Prudence M. Rice
Department of Anthropology
Southern Illinois University
Carbondale, IL 62901

The symposium on prehistoric and historic ceramic kilns at the 98th annual meeting of The American Ceramic Society in 1996 had its origins in a number of earlier occurrences. One was finding myself supervising excavation of a buried adobe structure that turned out to be an early 17th-century Spanish colonial kiln in Moquegua, in far southern Peru (Rice and Van Beck 1993; Rice 1994). Excavations at this kiln took place over three summer field seasons in the late 1980s, and during the periods between fieldwork I tried to research the literature on prehistoric kilns, kiln structures and variability, and excavation of kilns. I was disappointed to find that surprisingly little was available on these topics that I could use as a guide for my own excavations and interpretations.

Another factor prompting this symposium was the writing of a review article on recent ceramic research in archaeology (Rice 1996), which led to the discovery of numerous references to prehispanic kilns at sites in the New World. Given the papers in this symposium, this is hardly a surprising revelation, except it was not too long ago that New World archaeologists were taught that precolumbian New World pottery was manufactured entirely without the wheel, without glazes, and without kilns. These technologies were thought to be absent from potting activities in the western hemisphere until their introduction by Europeans. The recent discoveries of kilns--often quite *early* kilns--at precolumbian sites in North, Central, and South America are obviously forcing us to discard that old notion. But the discription of these discoveries is often scattered among specialist publications and field reports with low circulation, and is not readily available to researchers in other culture areas.

This is what led me to contact Prof. W. David Kingery and propose a symposium on kiln technologies past and present, to be held at The American Ceramic Society meetings. I suspect that more New World archaeologists might better recover evidence of kilns and pyrotechnologies if they had a clearer idea of what to anticipate--how to recognize these structures when abandoned and buried (the kiln interpretation is rarely immediately evident: see Redman and Myers 1981; Rice 1994; Feinman and Balkansky, this volume), what kind of variability exists in kilns, and how to excavate them. That knowledge is not a customary part of the received wisdom and training in New World archaeology, in part because prehistoric archaeologists traditionally have not expected to find kilns. But also, as noted, the most recent findings are not easily accessible at present. I thought it would be useful to provide a forum bringing together archaeologists working in both the Old and New Worlds, and also ceramic engineers with expertise in ceramic history and pyrotechnologies.

The subject of kilns is many-faceted, and in order to limit our subject to a one-day symposium I was forced to focus on only a few of these facets and, regretfully, exclude others. In assembling the contributions, I hoped to see some of the following issues addressed: social and economic questions relating to kiln design, construction, maintenance, and repair; technical and economic issues of site selection, function, fuel use, and temperatures attained; questions of output and volume of production; origins and spread of kiln-firing technology; relations with other pyrotechnologies; and so forth. Readers interested in archaeological pottery and firing technologies will notice a striking omission in this want list of topics: experimental and ethnoarchaeological appproaches to firing. As can be seen in the papers, however, many participants in the symposium have incorporated such approaches into their own research and teaching, and indeed the final hour of the symposium was a round-table discussion of the role of such studies. But formally including ethnoarchaeological studies of pottery firing in the symposium agenda would have expanded it into a multi-day enterprise, and I elected to focus more on the kilns themselves.

What IS a kiln? Just as the question, "What is a clay?" may be answered differently by mineralogists, soil scientists, potters, and archaeologists, so kilns may be defined differently by engineers, potters, and archaeologists. This seemingly innocuous question turns out to be especially problematic for archaeologists. Part of the difficulty is that they must wrestle with little more than fragmentary remains in the field, and some kilns are nearly invisible on the surface; another problem is that archaeologists must interpret a long developmental history of prehistoric kiln forms that cannot always be easily pigeonholed into neat typological categories.

For example, we can consider the definitions and classifications of kilns as set out by the modern ceramics industry. According to the *Dictionary of Ceramics* (Dodd and Murfin 1994:176), a kiln is a "high-temperature installation used for firing ceramic ware or for calcining or sintering." The entry goes on to note different kinds of kilns for the two categories of products (readers should note that many of the terms and definitions in this source have a decidedly British rather than American slant). Many variants of standard modern kiln types bear the names of the individuals or companies that developed them and/or own the patents. Even different types of kiln settings may be named.

It is also appropriate to note here that the term "furnace" is often used interchangeably with "kiln," although "furnace" more typically designates a very high temperature apparatus used in making glass (see paper by Patrick McCray, this volume), melting metals, and smelting ores. The *Ceramic Glossary 1984* published by The American Ceramic Society does not define either kiln or furnace, but instead identifies types of these installations. Kiln types include: "chamber; circular; clamp; downdraft; dry; Hoffman; muffle; periodic; rectangular; rotary; round; scove; tunnel" (Perkins 1984:47). Types of furnaces, on the other hand, include: "box; continuous; continuous tank; direct-fired; end-fired; end-port; pot; recuperative; regenerative; semimuffle; side-fired; side-port; tube; U-type" (Perkins 1984:35).

Archaeologists are unlikely to be familiar with technical definitions of the kiln types used in the modern ceramics (including brick and glass) industries as discussed in some of the papers in this volume, and industrial scientists may be unfamiliar with the range of variability of "primitive" and often fragmentary kilns used in prehistoric times around the world. For these reasons, it is worthwhile here to review some of those technical terms and definitions. Note that, apart from the major categories of kilns based on operating characteristics--e.g., intermittent and continuous, updraft and downdraft--types are not always mutually exclusive.

TYPES OF KILNS

I. Calcination or sintering kilns - these are used to produce changes in the physical structure of various minerals by means of the application of heat, for example through removal of water; e.g., calcination of gypsum or calcite prior to manufacture of plaster. Types of modern calcination kilns include shaft kilns, rotary kilns, and multiple-hearth furnaces (Dodd and Murfin 1994:176). In the Spanish colonial world, where some calcination kilns have been excavated; these were simple shaft (updraft) kilns similar to those used for pottery (see Rice and Van Beck 1993; Eicholz 1983; Costello 1977; for discussion and a non-Spanish example, see Wheaton, Reed, and Gantt 1987).

II. Ceramic kilns - these are used for firing pottery and ceramic materials such as earthenware and terracotta (including brick and floor tiles), stoneware, porcelain, etc. Ceramic kilns are described as being either intermittent (also called periodic) or continuous kilns, and several variants of these two major types are briefly described below (see Dodd and Murfin 1994); in addition, the chapter by O.J. Whittemore describes some other unusual constructions.

A. *Continuous kiln* - "a kiln in which the full firing temperature is continuously maintained in one or other zone of the kiln" (p. 72). Continuous kilns are mechanized installations commonly used in 20th-century ceramic industries for firing tablewares as well as heavy clay products (bricks, tiles, etc.). In these kilns, carloads of stacked wares are physically moved through a series of chambers to be preheated, fired, and then cooled. This, plus other types of automation, has greatly speeded up the manufacturing process by eliminating human handling.

1. *Annular kiln* - "a large continuous kiln, rectangular in plan, of a type much used in the firing of building bricks" (p. 13).

- Longitudinal-arch kiln - an annular kiln "in which the axis of the arched roof ... is parallel to the length of the kiln" (p. 189)
- Transverse-arch kiln (or continuous chamber kiln) - an annular kiln "divided into a series of chambers by fixed walls .... The axis of each chamber is transverse to the length of the kiln" (p. 335).

2. *Tunnel kiln* - "a continuous kiln ... in which ware passes through a stationary firing zone near the centre of the kiln" (p. 339). Commonly a carrying car loaded with brick travels through a long straight tunnel.

3. *Orbit [or rotary] kiln* - "a continuous kiln for firing tableware. Each item is fired separately, and is mounted on an alloy support ... which rotates the ware on its own axis as it passes through the kiln, thus ensuring uniform temperature distribution" (p. 221).

B. *Intermittent (periodic) kiln* - the type most familiar to potters and archaeologists, intermittent kilns are "batch" kilns, in which the kiln is loaded with green ware, fired, cooled, and then unloaded (see p. 167).

1. *Downdraft kiln* - "A kiln in which the hot gases from the fireboxes first rise to the roof, then descend through the setting and are finally withdrawn through flues in the kiln floor," exiting via a chimney (p. 97). The multichambered "climbing kilns" of China are examples of this type (see Whittemore, this volume).

2. *Updraft kiln* - "An intermittent kiln in which the combustion gases pass from the fireboxes through the setting and thence through one or more chimneys in the roof. Such kilns are inefficient and are now obsolete" (p. 342). In terms of the position of the flame relative to the ware, updraft kilns may be "vertical" (single or double-/multiple-chambered) or "horizontal."

Other kiln types, variants or combinations of some of the above types and features, are named and described more specifically:

- One of the best known of the modern periodic kiln types is the Hoffman ring kiln, which may be described as a top-fired annular tunnel kiln with transverse arches, used for firing bricks (p. 332). Whittemore (this volume) describes Hoffman ring kilns in Seattle and Brazil; I have seen them in use in Shandong, China (coal-fired), and on the Bolivian *altiplano*.
- Open-flame kiln - "A kiln in which the ware being fired is exposed directly to the burning gases of the flame" (p. 220). This term would apply to pit kilns and single-chamber updraft kilns, for example.
- Clamp - "A kiln constructed, except for the permanent foundations, of the bricks that are to be fired, together with combustible refuse ..." (p. 63). Clamp kilns, and the virtually identical "scove kiln," are updraft firing. For a modern example, see the paper by Julia Costello in this volume.
- Muffle kiln - a kiln incorporating a "refractory internal shell," which protects the ware "against contact with the combusion gases" (p. 210). Modern kilns/furnaces used in scientific laboratories are usually muffle furnaces.

- Round (beehive) kiln - "An intermittent kiln, circular in plan, with fireboxes arranged around the circumference. Such kilns find use in the firing of blue engineering bricks, pipes, some refractory bricks, etc." (p. 267). For an example from Venice, see the chapter by McCray.
- Rectangular kiln - "An intermittent kiln, rectangular in plan, with fireboxes at intervals along the two long sides. Such kilns find limited use in the heavy-clay and refractories industries" (p. 255).
- Bottle kiln - "A type of intermittent kiln, usually coal-fired, formerly used in the firing of pottery; such a kiln was surrounded by a tall brick hovel or cone, of typical bottle shape" (p. 36).

The point here is that kilns past and present can be defined and classified on a number of different criteria, including (but not limited to): continuity of firing; permanence of the structure; products or kinds of wares fired; position of wares relative to heat source; number of chambers; shape or plan; kind of fuel used; and so forth. How useful are these classifications in historical perspective? Reasonably useful, because they underscore the importance of many of the characteristics that were important in the evolution of firing technology, characteristics that find their expression, one way or another, in today's kilns used by industry and by studio potters. How useful are these classifications for prehistorians and archaeologists? Somewhat less so, perhaps, because the emphasis is not consistently on the kinds of physical traces that survive millennia of weathering, site reoccupation, and other destructive forces.

PAPERS IN THIS VOLUME

The papers assembled here explore some of this variability of kilns past and present. Organization of the volume is partly chronological and partly geographical, and stresses the rather revolutionary new data on kilns to emerge from recent archaeological excavations at sites in North, Central, and South America. The geographical and temporal areas covered include Neolithic Greece, 3rd-millenium Indus valley, imperial China, the southwestern United States, coastal Peru, Classic-period (ca. AD 300-900) Mesoamerica (Mexico and Honduras), historic period California, and Renaissance Italy. Production contexts range from small-scale household production to nucleated industries.

The first contribution comes from co-organizer David Kingery, who begins, fittingly, at the very beginning: Kingery observes that the excavation of archaeological kilns is, in some senses, the reverse of the original process of designing and constructing them. In order to understand the functioning of kilns, he reasons, we need to start with comprehension of the basic "operational principles" that underlie the construction and operation of all kilns, from the simplest to the most sophisticated. Four critical functions are at the core: generating heat, transferring heat, containing heat, and providing a stable setting for the ware being fired. His contribution highlights the many ways these basic design principles have been implemented around the world, from earliest times up through modern industrialized production.

Next, Karen Vitelli discusses firing of the earliest pottery represented in this volume. She argues that some Early and Middle Neolithic pottery from southern Greece was kiln-fired. By means of painstaking analysis of the technical characteristics of the pottery wares, combined with experimental constructions and firings carried out with her students, she arrives at an unusual and provocative suggestion: use of "portable kilns." Basically large bottomless clay funnels or cylinders (miniature bottle kilns), these "kilns" would have served three of the four functions outlined by Kingery: containing and transferring heat, and containing the stacked ware; they also may have held the fire itself. These Neolithic "kiln" constructions suggest that early potters carefully observed the successes and failures of their firing efforts and incorporated their findings to improve their technology and their products. The complex and sophisticated firing strategies employed by Greek potters are perhaps best known through the three-stage firings used for the beautiful black and red figured pottery of later Classical times, but those accomplishments clearly culminate a history reaching back five millennia or more. Vitelli's suggestion should prompt archaeologists working in other areas to look more closely at their sherds and think creatively about how early potters might have designed innovative approaches to the critical firing process.

Heather Miller reviews past and recent archaeological literature on the Harappan civilization of the Indus valley (Pakistan and India) in order to describe the range of firing structures and technologies of the 3rd millennium B.C. A variety of firing strategies was in use, including "open" (but not bonfire) firings, single-chamber ("pit" or "oven") structures, and double-chamber updraft kilns, for producing terracotta vessels and other objects. Miller calls attention to the existence of a gradation of types of firing constructions and structural features (walls, roofing), marked by increasing complexity and permanence. However, she emphasizes that this continuum is not an evolutionary sequence. Rather, all these structures were in more or less contemporaneous usage, calling attention to the need for further research to answer questions about why Indus craftspeople used certain kinds of kilns for particular products. Her review of the literature emphasizes the need for archaeologists to provide more detailed information on possible kiln structures in their reports, and she provides a useful outline of the kinds of data needed.

The contribution by Hu and Li on "egg-shaped" kilns documents a historically important variant of an intermittent or periodic downdraft kiln in Jingdezhen, China, one of the premier centers of porcelain production for many centuries. This traditional kiln, fired by pine wood, is unique because of the thermal behavior conferred by its distinctive shape and construction, with the firebox in front. Specifically, the velocity of flow of hot gases is increased from top-to-bottom and front-to-rear of the kiln by virtue of the constriction of the firing chamber, thereby enhancing convection heat. The insulating air layer around the walls also helps achieve temperature uniformity. The "egg-shaped" kiln illustrates--indeed optimizes--all four of Kingery's principles for generating, transferring, and containing heat, while also providing a stable setting for the ware. The setting of saggars is singularly important in this kiln type, in that it allows products with different maturing temperatures to be fired simultaneously.

In the first of five papers on precolumbian kilns in the New World, Eric Blinman and Clint Swink, archaeologist and potter respectively, summarize the findings of recent studies of trench kilns in the Anasazi area of the Southwestern United States. Dating primarily between A.D. 900 and 1300, these kilns are shallow rectangular pits, often occurring in groups, that appear to have been used multiple times. Several years of experimental firings in kilns of this type, carried out as a collaborative effort among archaeologists, potters, and Native Americans, have transformed our understanding of how the beautiful Anasazi black-on-white pottery was produced. The authors surmise that the kilns were used in collaborative firings of the work of multiple potters, given the size of the kilns and the different decorative styles, paints, and pastes of wasters recovered from the trenches. The location of the kilns, well away from residences, suggests that firing areas may have been chosen largely on the basis of fuel abundance and quality.

Izumi Shimada's contribution begins with the common assumption held by archaeologists that prehispanic Andean pots were pit fired, and then he proceeds to review ceramic firing structures in coastal Peru, particularly north coastal Peru. This area displays an array of prehistoric kiln types, including circular pit kilns, semi-closed, double-chambered horizontal pit kilns with chimneys, rectangular kilns with basal walls on two sides, and rectangular kilns with masonry walls all around. Much recent work on kilns has been done by Shimada himself at two sites. One of these is Batán Grande, where at least 57 kilns were found in a small area. These kilns, often occurring in clusters, are semi-enclosed updraft pit kilns with domed roofs and chimneys, and exhibit a surprising amount of variation (see also Shimada et al. 1994). At another site, Pampa Grande, he found a multi-roomed pottery-making workshop with a small, rectangular masonry kiln. Shimada concludes that this kiln type may represent "a relatively clean, space-saving solution to ceramic production in urban settings" (see also Arnold 1991) as well as affording "close supervision over the production of high quality ceramics decorated with religious or political symbolism." Shimada's research argues convincingly in favor of multi-stage, multidisciplinary studies of ceramic production involving archaeological excavations, replicative experiments of forming and firing, and archaeometric analyses.

Gary Feinman and Andrew Balkansky discuss modern village/household-based pottery production in the valley of Oaxaca, in south-central highland Mexico, as a basis for comparison with the prehistoric situation. A variety of firing arrangements can be found in this contemporary setting, some of which echo the findings of archaeological investigations. More than 20 years ago, excavations at the large site of Monte Alban, in the central valley, led to identification of updraft kilns, and these (Winter and Payne 1976) represent one of the earliest references to kilns in the prehispanic New World. Recent excavations at Ejutla in the southern part of the valley recovered evidence of pottery firing activities in a residential setting in the form of pit kilns excavated into bedrock. Firing of an experimental pit kiln, constructed along the same design, left few vestiges that would be archaeologically recognizable. The authors conclude that pit kilns could have been much more widespread in prehistory, but problems of archaeological preservation and research strategies often impede recognition of these features.

Christopher Pool summarizes findings from excavations at the site of Matacapan, in southern Veracruz, Mexico, where pottery was fired in small, circular, updraft kilns after about AD 300 (and perhaps as early as 100 BC). As do several other contributors to this volume, Pool notes that the prehispanic kilns are similar to ones used in modern times and that at Matacapan kiln firings were used simultaneously with open firings. However, different types of firing strategies were used for the four wares produced at the site and for different modes of production organization, household vs. "industry." The most important factor influencing the use of kilns seems to be better control over the rate of heating, especially for fine-paste wares; one coarse ware was evidently fired in the open. Control of the firing atmosphere was also a sigificant factor, differentiating Fine Orange and Fine Gray wares.

Patricia Urban, Christian Wells, and Marne Ausec provide evidence for probable kiln structures in the Naco valley of Honduras, lying on the "southeastern periphery" of the Maya lowlands. Excavations at the site of La Sierra, the capital of the region, revealed pottery manufacturing areas in residential zones lying outside the civic-ceremonial site core. Here they found borrow pits presumably used for excavating and slaking clay, workshop structures, pottery-making tools and resources, storage facilities, and stone foundations of circular updraft kilns. Several other likely firing areas also have been found elsewhere in the valley. Their discussion of surface-collected vs. excavated indicators of pottery production is a sobering reminder of the difficulties of recovering unambiguous data on prehistoric economic activities.

While the five previous papers discussed prehispanic kilns in various parts of the Americas, Julia Costello discusses the brick and tile kilns introduced to Spanish colonial California and Guatemala. The brick and tile industry of Spain, Roman-period Hispania, had its origins in that ancient civilization and was brought to the New World in the early 16th century. In highland Guatemala, for example, the Spaniards introduced brick and tile making to supply the new capital city (modern-day Antigua). The Guatemalan industry remains virtually unchanged in the town of Tejar and nearby places in the Chimaltenango valley. Costello used a strategy that combined analysis of published reports on the Roman industry, observations and interviews with tilemakers in Tejar, and documentary sources on the 18th-century Spanish missions in California, to interpret the archaeologically excavated remains of a tile kiln at Mission San Antonio de Pádua, California.

The Roman influence is also seen in the post-medieval European glass industry, which Patrick McCray explores by way of a review of glass making in

Renaissance Venice. After the fall of Rome, glass making diverged into two separate traditions, northern and southern (Mediterranean), the latter being characterized by an urban focus, different resources, and round or "beehive"-shaped furnaces. The Italian glass industry can be investigated through contemporaneous written sources, which reveal that everything needed for the Venetian glass industry had to be imported, including materials for constructing the furnace, quartz, clays for the crucibles, fuel, pigments, and so forth. McCray reviews the documentary sources and illustrations, adding chemical, SEM, and petrographic analyses of crucible fragments discovered in archaeological contexts. Venice's glass industry was enormously complex not only in its infrastructure (supply of raw materials), but also in terms of its facilities (for example, various kinds of furnaces or parts of furnaces for fritting, melting, annealing) and teams of technical specialists who had to carefully orchestrate their efforts in production.

And in the final contribution to the volume, O.J. Whittemore describes seven kilns, "unusual" in their construction as well as in their operation; some of them he himself fired, so he brings a personal element into the discussion. All these installations demonstrate the firing principles that David Kingery outlined and illustrate the varying ways in which they can be incorporated into kiln design.

REFERENCES

Arnold, P.
1991 *Domestic Ceramic Production and Spatial Organization: A Mexican Case Study in Ethnoarchaeology.* Cambridge University Press, Cambridge.

Costello, J.
1977 Lime processing in Spanish California, with special reference to Santa Barbara. *Pacific Coast Archaeological Society Quarterly* 13(3):22-32.

Dodd, Arthur, and David Murfin
1994 *Dictionary of Ceramics* (3rd ed.). The Institute of Materials, London.

Eicholz, J.C.
1983 A Spanish colonial lime kiln. *Proceedings of the Ninth International Congress for the Study of the Pre-Columbian Cultures of the Lesser Antilles, Santo Domingo, 2-8 August, 1981.* Centre de Recherches Caraibes, Université de Montreal, Montreal.

Perkins, W.W. [ed.]
1984 *Ceramic Glossary 1984.* The American Ceramic Society

Redman, C.L., and J.E. Myers
1981 Interpretation, classification, and ceramic production: A medieval North African case study. In, *Production and Distribution: A Ceramic Viewpoint,* H. Howard and E. Morris, eds. pp. 285-307. B.A.R. International Series 120, Oxford.

Rice, P.M.
1994 The kilns of Moquegua, Peru: Technology, excavations, and functions. *Journal of Field Archaeology* 21(3):325-344.

1996 Recent Ceramic Analysis: 2. Composition, Production, and Theory. *Journal of Archaeological Research* 4(3):165-202.

Rice, P.M., and S.L. Van Beck

1993 The Spanish Colonial kiln tradition of Moquegua, Peru. *Historical Archaeology* 27:65-81.

Shimada, I., C.G. Elera, V. Chang., et al.

1994 Hornos y producción de cerámica durante el período Formativo en Batán Grande, costa norte del Perú. In, *Tecnlogía y organización de la producción de cerámica prehispánica en los Andes*, I. Shimada, ed., pp. 67-119. Pontificia Universidad Católica del Perú, Lima.

Wheaton, T.R., Jr., M.B. Reed, and M.E. Gantt

1987 *The Jimmie Green Lime Kiln Site, Berkeley County, South Carolina.* Garrow & Associates, Inc., Atlanta.

Winter, M. C., and W.O. Payne

1976 Hornos para cerámica hallados en Monte Albán. *Boletín del Consejo de Arqueología* 16:47-40. INAH, Mexico City.

OPERATIONAL PRINCIPLES OF CERAMIC KILNS

W. David Kingery, Regents Professor
University of Arizona
Tucson, Arizona

The problem of recognizing archaeological kilns and devising inferences about their operation is the reverse of the problem of designing and constructing these kilns. Understanding the design process is a necessary first step toward interpreting the archaeological record. We shall begin at the beginning.

As engineers or craftspersons approach a design problem, their first requirement is to have a conception of the operational principles involved. The operational principle of a machine or device has been formulated by Michael Polanyi (1962:328) as "How its characteristic parts fulfill their special function in combination to an overall operation which achieves the purpose." That is, how it works. It may well be that the craftspeople designing prehistoric kilns would use entirely different descriptors for the operational principles than we would. Nevertheless, to achieve a successful design tested by empirical performance sufficient to develop into a normal configuration observed in the archaeological record, they would have put these principles into practice. The operational principle of a ceramic kiln is most broadly to raise the temperature and thereby solidify ware by generating heat and transferring this heat to the ware. This is accomplished by

- Heat generation, most often by combustion
- Heat transfer to the ware
- Heat containment
- Providing a stable setting for the ware

How each of these functions is achieved defines the nature of the kiln and the firing process used.

Maniatis and Tite (1981:59) investigated the fired structure of more than a hundred prehistoric ceramics with scanning electron microscopy. They found that about half showed no evidence of vitrification at a magnification of 10,000X. It is convenient to divide the wide spectrum of ceramic wares into the following categories

- Non-vitrified ware, typically fired at 500-800°C.
- Some vitrification, typically fired at 800-1050°C.
- Fully vitrified stonewares and porcelains, typically fired at 1100-1300°C.

For non-vitrified wares the hardening process results from a breakup of the crystalline clay structure to form an amorphous phase which agglomerates and sinters by surface diffusion (Kingery 1992). The result is shown at high magnification for a prehistoric sample from Halaf.

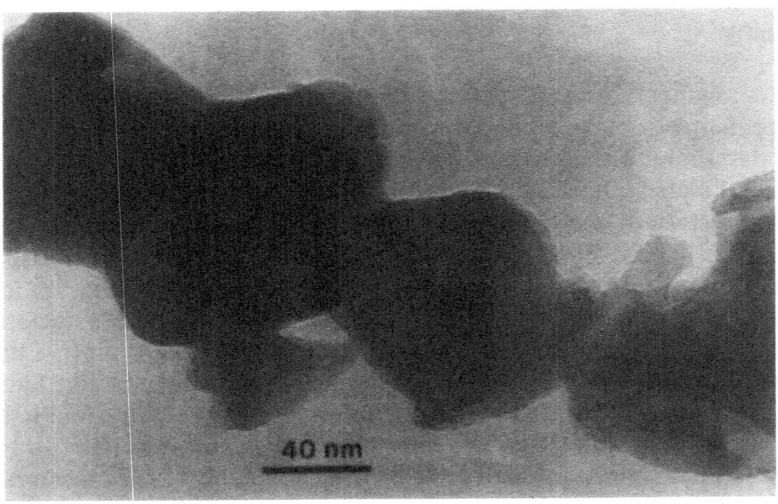

Figure 1. Transmission electron microstructure of clay particles sintered together as a result of low temperature firing.

These processes are rapid and do not require a long firing time. Importantly, the firing shrinkage is small so that rapid heating does not cause difficulties. Setting of the ware can be haphazard with no special arrangements to accommodate the low firing shrinkage and non-sticking of the pieces together.

For these wares open firing systems work very well. The source of heat is rapidly burning straw, sticks of wood, dung or any other combustible material distributed within, below and surrounding the ware as shown in Figure 2.

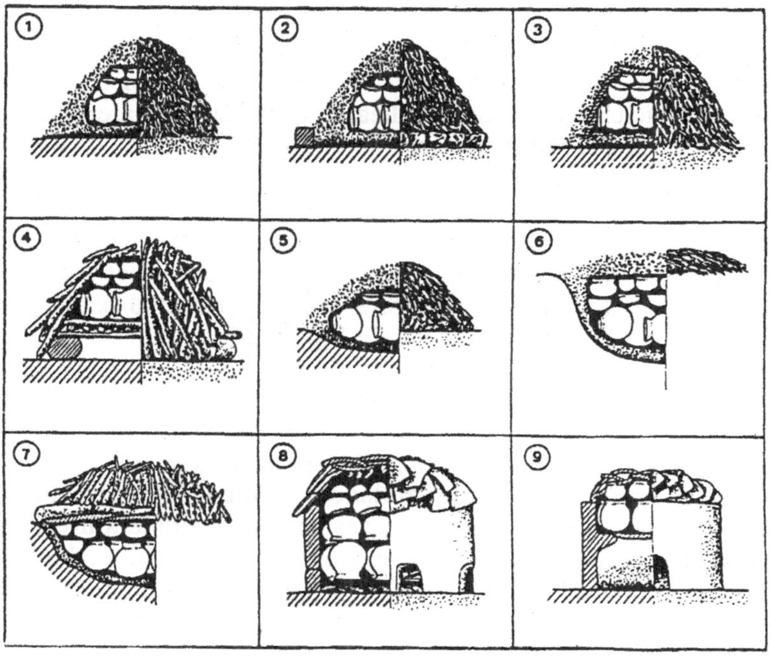

Fig 2. The main firing structures recorded by Gosselain and Smith (1995) in Sub-Saharan Africa.

Heat is transferred to the ware by convection and radiation from the hot burning fuel. In some cases, as shown in Figure 2-4, the setting is elevated to allow combustion air to enter below increasing the rate of convection and providing insulation of the ware from the earth below. Ware is supported by other ware stacked in an arrangement with open spaces between the individual pieces to accommodate heat transfer. Heat is maintained within the setting by piling up fuel above as well as within and below the ware. Burning fuel above the ware acts as a blanket enclosing heat and providing a residual layer of highly effective insulating ash. As will be clear from Figure 2, the residual traces of these bonfirings are meager. Areas of heated ground and ash may result from these or from other activities; such bonfire kilns can probably best be identified by association with other ceramic processing material.

For wares fired to achieve some vitrification a longer firing temperature is required. This is achieved by more effectively containing the heat in a permanent enclosure, such as illustrated in Figure 2-8 and 2-9. An intermediate arrangement is shown in Figure 2-7 where a pit is used with combustibles under the ware and a layer of fireproof material, almost always sherds, covering the ware to create a temporary enclosed structure. In permanent structures the source of heat is combustible material which can be inserted at the base of the structure or burned below it as may have been the case for oil firings. Ware is stacked on ware in a configuration that provides openings between the ware to allow hot gas flow and convective heat transfer. The top of the kiln is enclosed with a temporary roof of pot sherds. In order to allow more effective and uniform heat treatment the firing area is separated from the ware with a perforated plate to form a primitive updraft kiln such as shown in Figure 2-9.

As the ethnographic studies of Gosselain and Smith (1995) illustrate, (Figure 3), the temperature achieved in these different sorts of bonfire and primitive kiln constructions are not so different as is the length of firing and temperature uniformity. The permanent walled kiln structure more effectively contains the heat and achieves perhaps a temperature a hundred degrees higher. The added heat capacity of the kiln structure takes up much of the heat produced and requires a slower increase in temperature as illustrated in Figure 4.

The capital investment in a permanent structure to contain the heat is warranted by the production of larger quantities of ware even if the temperature is not increased. Larger kiln size lowers the surface area for heat loss relative to the volume available for firing and thus improves efficiency. Larger kilns also allow more elaborate systems for setting ware, particularly if glazes are used which stick to one another on contact. The result of these considerations is the normal configuration of an updraft kiln illustrated in Figure 5 where the firebox and setting area are separated. A large firebox is required when bulky fuels are employed. In the simplest design the top of the kiln enclosure is formed by temporary addition of broken sherds forming a permeable layer rather than a permanent roof. A more sophisticated design has a permanent domed roof, a separate setting area behind the firebox and uses saggers to contain the ware and protect it from contact with the long flames required for effective heat transfer. Adjustable flue openings allow control of the hot gas flow which transfers heat to the ware and are used to adjust the temperature and temperature uniformity. This kind of arrangement has a separate firing box for heat

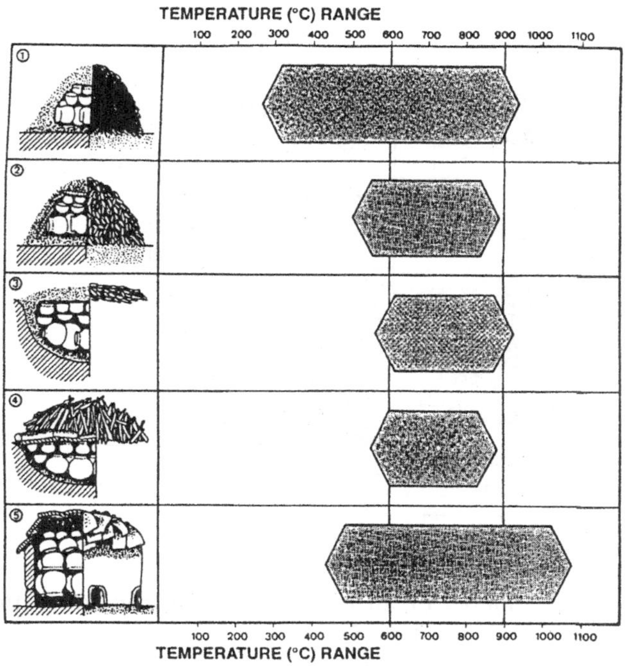

Figure 3. The temperature ranges observed ethnographically in five kinds of firing structures based on ethnoarchaeological studies in Sub-Saharan Africa (Gosselain & Smith 1995)

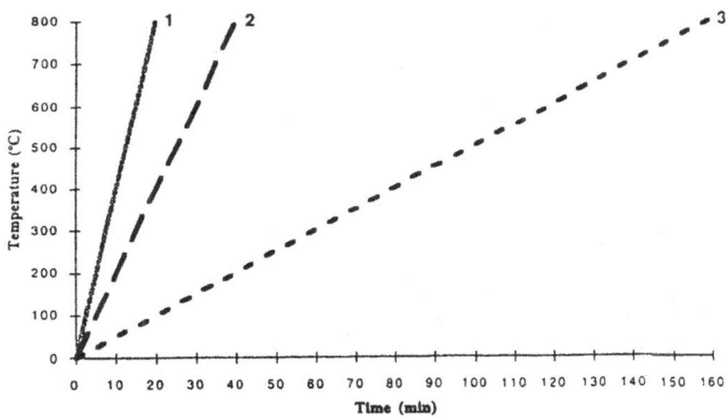

Figure 4. Estimations of heating rates for (1) open firing, (2) open firing with sherds separating pots from fuel, and (3) firing in permanent kiln structures (Gosselain & Smith 1995)

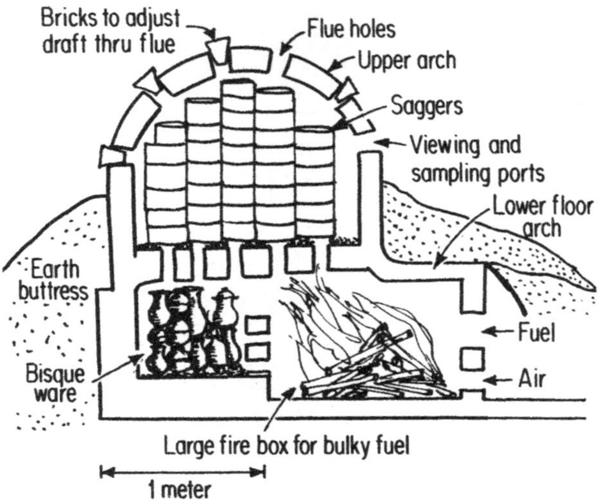

Figure 5. Typical components of an historic updraft kiln for pottery firing (Kingery and Vandiver 1986:255)

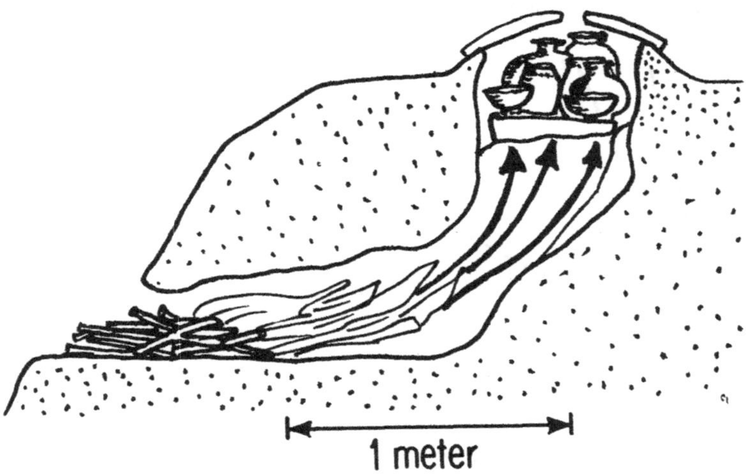

Figure 6. Underground kiln construction used in neolithic China (Kingery and Vandiver 1986:254)

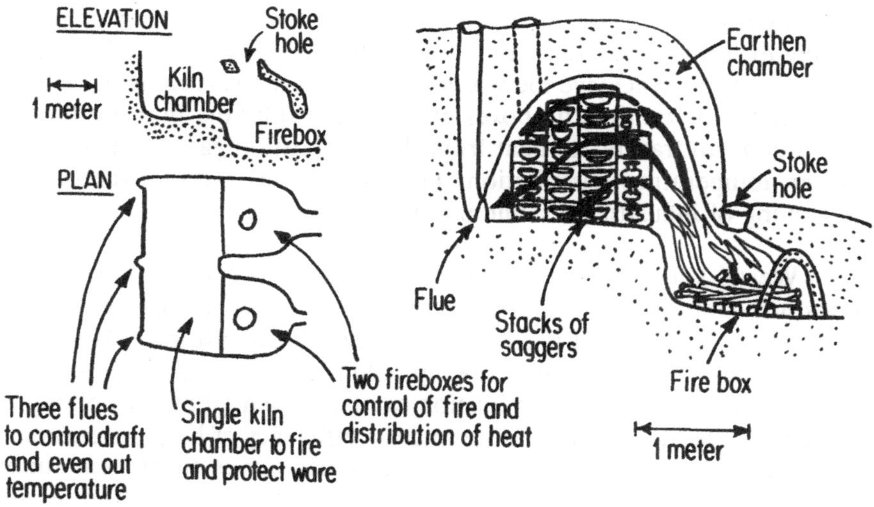

Figure 7. An underground stoneware kiln with separate fireboxes for heat generation and separate flues for gas flow control used in Song Dynasty northern China (Kingery and Vandiver 1986:98)

generation, updraft hot gas flow for heat transfer to the ware, a permanent wall and roof enclosure for the heat and a wide variety of methods for setting the ware that is most typical for historic ceramic production in the Western world.

A different tradition arose in China where sandy clay refractory soils allowed digging a kiln out of a hillside as shown in Figure 6. Kilns like this were used in early Neolithic communities; the best known example is at the Touristic Archaeological Site of Ban Po. The large heat capacity of the earth surrounding the kiln requires a long firing time. The evolution of this type of kiln allowed the achievement of high temperatures necessary for stoneware and porcelain production. A Song Dynasty kiln with two firing boxes for rapid heat generation, and a primitive fire wall for directing the flow of heat through the ware in a crossdraft/downdraft arrangement with multiple flues dug in the earth to adjust the hot gas flow, velocity and distribution to provide effective heat transfer to the ware is shown in Figure 7. The organizational elements of this kiln - heat generation, heat transfer to the ware, heat containment, ware support - determine its design and use.

Elaboration of this design for production of large amounts of ware on a much greater scale led to the development of the kilns at Jingdezhen described in a subsequent essay. In Europe, the production of porcelain during experiments at Dresden early in the 18th century was first thwarted by the inability to achieve the necessary temperatures. A rather simple cross-draft kiln with multiple fireboxes provided the solution. In order to achieve more uniform transfer of heat from the combustion gas to the ware, and a better temperature distribution, downdraft kilns were invented at the beginning of industrial ceramic production in large kilns.

We all know that some 80% of the air we breathe is nitrogen which does not contribute to our respiratory effectiveness nor to the combustion process. Instead it must itself be heated at the expense of heat generation for the ware. By moving the fire up a hill, for draft, along a long tunnel (dragon) kiln, the air for combustion can be preheated by the cooling ware, increasing the flame temperature and the efficiency of heat generation. The long path of the cooling combustion products serves to preheat the ware above the firing chamber, increasing the efficiency of heat transfer to the ware. In primitive kilns firing begins at the bottom of the tunnel and is moved up the hill as each section reaches the desired firing temperature. From this kiln design, it is only a small step forward to the modern tunnel kiln in which the fire remains fixed in position and the ware is moved on kiln cars or continuous rollers or belts past the firing point, preheating the combustion air on cooling and being preheated by the combustion gases before entering the firing zone.

By combining the idea of a dragon tunnel kiln with a downdraft kiln for temperature uniformity we reach the conception of a chamber kiln in which a tunnel is divided into chambers, each with a separate firebox and fire wall. Such kilns combine the effectiveness of air preheating and ware preheating with a downdraft temperature distribution in the kiln which achieves more effective heat transfer to the ware and uniform temperature distribution. This type of kiln was used during the Ming Dynasty in China. By connecting the ends of the kiln together in a circle one can eliminate the influence of the beginning and ending of the firing structure and produce a continuous chamber kiln which one still sees used today for brick and tile production in India, Pakistan, China and elsewhere.

In modern ceramic kilns, one sees all the types of kilns we have illustrated, plus more complicated industrial kilns for large amounts of ware designed for efficient heat generation, sometimes electrical, more often gas, in a variety of configurations. Advances have been made mostly in the combustion devices, which replace traditional fireboxes, in the arrangements for preheating combustion air and preheating ceramic ware, and most importantly, in the insulation for heat containment.

While there are an enormous number of individual ceramic kiln designs in each of these constructions the operational principle of hardening ware to produce useful properties is accompanied by

- Heat generation, most often by combustion
- Heat transfer to the ware, raising its temperature
- Heat containment
- Providing a stable setting for the ware during firing

By keeping these components of the most simple and the most sophisticated kilns in mind, sound inferences can be developed in interpreting the archaeological and historical records.

References

Gosselain, O. P., and Smith, A. L., 1995, "The Ceramic and Society Project: An Ethnographic and Experimental Approach to Technological Choices" KVHAA Konferenser, 34, 147-160.

Kingery, W.D., and Vandiver, P.B., 1986, Ceramic Masterpieces, Free Press, NY.

Kingery, W.D., 1992, "The Nature of Changes Occurring during Low Temperature Sintering, Ceramics: Adding the Value, M.J. Bannister, ed. CSIRO Publications, Melbourne, Australia, 1-10.

Maniatis, Y. & Tite, M.S., 1981, Jour. Arch. Sci., 8, 59.

Polanyi, M., 1962, Personal Knowledge, Univ. Chicago Press, Chicago, 328.

INFERRING FIRING PROCEDURES FROM SHERDS: EARLY GREEK KILNS

Karen D. Vitelli
Department of Anthropology
Indiana University
408 N. Indiana Avenue
Bloomington, IN 47408

ABSTRACT

No certain direct evidence has been found to date for the firing places and practices of the southern Greek Middle Neolithic potters (ca. 5500 B.C., calibrated) who manufactured a beautiful, highly sophisticated ware. Nevertheless, the high quality and consistency of the pottery and the considerable quantities that have been recovered (more than a ton from the Franchthi Cave excavations alone, and the ware occurs as the major component of all contemporary southern Greek sites) make it possible to suggest that the pieces were fired in a kiln, perhaps a portable version whose remains we may have in what have usually been considered fragments of large storage jars. The paper discusses the evidence for firing practices that can be extracted from Greek Early and Middle Neolithic sherds and attempts at firing in experimental kilns developed from that evidence. Among the criteria that suggest the reconstruction are hardness, color patterns--including the presence and location, or absence, of firing clouds and circles--condition of calcium carbonate inclusions, analyses of firing temperatures, degree of sintering of the surface slips, the relative sizes and shapes of reconstructed vessels, and details of potential kiln pieces.

THE ARCHAEOLOGICAL SETTING

From the most recently excavated and fully documented Neolithic site in southern Greece, Franchthi Cave, we recovered ca. two and a quarter metric tons of potsherds that span the first 3000 years of pottery production in Greece. For the archaeologist studying the material, that is a lot of pottery, but it averages out to less than 1 kg of pottery per year of occupation. Somewhat more refined calculations provide an impression of the probable scale of production in the Neolithic. The evidence from southern Greek sites suggests that, in the earlier phases that are the subject of this paper, exchange of ceramics among sites was very limited. Most of the pottery found at Franchthi was probably made, used, and discarded there. Thus, if we adjust for the average weight of vessels, the percentage of the site excavated, and the duration of each phase, we may use the weights of recovered sherds to estimate annual production at about 12-13 vessels in the earliest phase, and roughly 150-160 during the phase of peak production (Vitelli 1993:210, 211 notes 9, 10). Even if those estimates are off by several magnitudes, we are looking at rather small scale production in the following discussion.

The earliest phase of production at Franchthi, the Early Neolithic, spans at least 400 years at the beginning of the 7th millenium B.C. (calibrated). The potsherds from this phase can be assigned to one of five categories based on the raw materials, or, more specifically, on the nonplastic inclusions, used for each. All five ceramic bodies were in contemporaneous production throughout the 400-year period, and at least four of them were probably produced locally, at Franchthi. All five wares were made in the same general range of sizes and shapes, and as far as we can tell, all shared similar properties of, e.g., porosity and thermal shock resistance. Their differences are in their initial raw materials, and in the ways in which they were built and fired (Vitelli 1993:95-116 passim).

Although we have not yet identified anywhere in southern Greece a place or feature that we can securely claim as a firing installation used by these earliest potters, the coloration of their pots gives us useful clues as to how they were fired, and sufficient evidence to recognize that different techniques and, possibly, fuels were used for each ware. The primary clues used in piecing together this puzzle are simple: fired clay colors are primarily dependent on the state of oxidation of iron oxides. Iron oxide-rich bodies generally fire reddish, or light-hued colors in the presence of an abundance of oxygen; black or dark gray if oxygen is scarce or the atmosphere is reducing, and yellows and lighter grays if oxidation or reduction is incomplete (Rice 1987:334-336). These colors do not tell the whole story, but they provide valuable starting points.[1]

EARLY NEOLITHIC FIRING PRACTICES

Lime Ware

I review here only the three Early Neolithic wares for which the sample size is largest and, hence, the evidence clearest. The category that was produced with the greatest frequency is Lime ware, so called because of the calcium carbonate inclusions, at least some of which are crushed and intentionally added calcite, up to 2 mm in maximum dimension. The sherds--and the rare reconstructable pot--are predominantly light tan/pink in color (Munsell 2.5 YR 5-7/4-8), but sherds are regularly mottled with irregular yellow, gray, and black areas of fire-clouding. The cores are generally the same color as the adjacent surface, i.e., uniformly light-colored, except where vessel surfaces were clouded gray, in which case the entire core is usually gray as well. Where vessel walls are particularly thick, as around base joints or at thickened rim folds, the core is likely to be gray at the very center. Well-preserved burnished surfaces show a hardness of 3-4 on Mohs' scale. Lime pops are very rare.

The essential absence of lime popping on pots tempered with sizable calcium carbonate fragments, as well as the hardness range, are consistent with exposure to temperatures not exceeding 700-800° C, temperatures readily reached in an unconfined wood fire. The firing clouds point to direct contact with fuel, possibly with other pots, and to changing atmospheres during firing. The dominance of light colors suggests a relatively clean burning fuel with a light ash that fell and collected in small patches to produce some of the clouds, but was not so firm as to cover whole pots or large areas of pots. The atmosphere was mainly oxidizing throughout the firing, with temperatures above 500-600° C maintained long enough to burn out the organic matter in the clay in all but the thickest areas. The pots were probably allowed to cool in place, in contact with other pots, ash and coals, which left the fire clouds that penetrated deep into the cores. Differences in the interior and exterior colors on individual vessels, as well as contrasts between rim and bottom colors, suggest that some pots were fired right side up, others, upside down. Nothing about the Lime ware assemblage insists that any special construction, such as containing walls or a pit, was used, but they remain a possibility.

Sandy Ware

Sandy ware is the second category, and was made from a clay body with substantial quartz sand nonplastics and virtually no calcium carbonate. These sherds are also often fire-clouded, but in this case, the clouds are lighter colors, contrasting with the predominantly dark brown, gray or black colors (Munsell 2.5-7.5 YR 4-5/1-4) of the bulk of the assemblage. Cores are generally dark, often deep black, rarely reddish brown. Hardness is slightly higher than that of Lime ware (Mohs' 3-5) although the pieces often "feel" harder and do not generally show the same degree of wear that is common on the Lime ware sherds.

Many, if not all of the Sandy pots appear to have been coated on the exterior and just inside the rim, with paint that occasionally dripped and splattered on the interior. This paint was probably not a clay slip, but ground from an iron-rich lump of mineral, since individual pigment grains are frequently large enough to be visible to the naked eye. Even in areas of a pot that have fired deep black, these red pigments grains are sometimes visible in unburnished depressions, and several examples of the pattern painted variety of Sandy ware show dramatic red stripes against the dark black ground of the fabric.

The dark surfaces and cores of Sandy ware sherds could result from the incomplete oxidation of carbonaceous clays, reduction, or smudging--the last involving deposition and absorption of carbon soot during the final cooling stages of the firing. Possibly, all three processes were involved at some point, but reduction seems to have been the dominant process, since, in theory, incomplete oxidation would have occasionally produced, on a section through the wall, a graded pattern of light subsurfaces and a darker gray at the center of the core, while soot deposition would have affected the red grains and painted lines as much as the more porous body (Shepard 1968:39). When the red pigment did not reduce to black, it was presumably because the larger grain size remained unaffected at the temperature range and soaking time that reduced the finer iron oxide particles in some of the paint and in the clay body. The original clay body may have been very high in organic content, contributing to the darkness of the cores (incompletely burned organic materials) and to the reduction atmosphere. Rye says that "reduction of iron oxide does not start below about 900° C" (Rye 1981:108). In experimental firings of various kinds with wood fuel we have found it relatively easy to reach 750-850° C in several hours, but it then becomes very difficult to go any higher, in part because the intense heat makes it practically impossible to approach the fire closely enough to add more fuel carefully. Either the Sandy ware potters achieved higher temperatures by using some kind of enclosure to retain the heat and maintain the reducing atmosphere around their fire, or the presence of fluxes and carbonaceous matter promoted reduction at somewhat lower temperatures. They must have allowed their pots to cool in place, covered with ashes or something that prevented the surfaces from oxidizing. Small currents of air frequently penetrated the cover, however, and caused the light-colored clouds. While a number of questions about the details of the Sandy ware firings remain, and the firing may well have involved the use of some purposeful enclosing structure, it is clear nevertheless that the firings were, consistently, different than those that fired the contemporary Lime ware.

Ungritted Ware

The potters who made the third Early Neolithic ware used still different procedures. The ware occurs in very low frequencies at Franchthi (less than 1% of the assemblage), but at Lerna and other sites a few days' walk to the north, it occurs with greater frequency, and in several varieties. Quite a few complete or nearly complete vessels have been restored from those sites. The pots were made from a highly calcareous clay body that contains essentially no nonplastic inclusions visible with a 10X hand lens, hence its name, Ungritted ware. Most of the sherds are now soft and powdery, probably the result of having been cleaned by the excavators in hydrochloric acid without rinsing in fresh water, a once-standard procedure at excavations in southern Greece where sherds often come from the ground encased in a thick calcareous crust (Vitelli 1993:11 note 2; Dowman 1970:117-118). Most Ungritted sherds are now easily scratched by Mohs' 1-2, but an occasional well-preserved sherd tests as high as Mohs' 5. Most sherds and vessels are a pale yellow-tan (Munsell 7.5 YR 7-8/4-6) with dark firing clouds on the exterior. Fewer, but still common, are those that are uniformly oxidized on the exterior to a yellow-tan color, although the base area is sometimes gray. Sometimes the cores and interior surfaces are the same light colors, but frequently the core and interior are a uniform pale, silvery gray.

Less common than either the mottled or the uniformly light pots, but still occurring in substantial numbers, are vessels that have fired to a uniform pale silvery gray color on all surfaces, which are slightly darker than the cores, and generally without firing clouds. Their color is comparable to other "gray wares" in the literature from Greece (in the Late Neolithic of Thessaly, and the Middle Bronze Age) and from other cultures, and is particularly interesting because it represents uniform and extended exposure to an atmosphere intermediate to reduction, which would have produced the dark gray or black seen in some of the firing clouds of the variegated pieces, and oxidation, which produced the yellow-tans. The pale gray is comparable to the color seen in the transition area of firing clouds, but that intermediate atmosphere is very difficult to control, much less maintain for the period of time necessary to affect both surfaces and the full core. The gray pots point to firing in a contained environment, where the pots were not in direct contact with the fuel and the atmosphere surrounding them was lacking in free oxygen and full of lightly reducing gases.

Since these vessels occur at the very beginning of pottery production in Greece, in all the color combinations noted, I suspect the results were initially accidental, a result of the way pieces were stacked in the fire, the kind of fuel, and the timing of its additions. All three varieties--the uniformly light-colored pieces (with light or gray interiors), the mottled pieces, and the uniformly pale gray pieces--could have come from a single firing.

The varied results could happen if a group of pots was stacked with, for example, the bottom pot right side up (Fig. 1, pot #1), a smaller pot placed inside it, (Fig.1, pot #2) and a third pot, upside down, resting over, or on the rim of the first pot (Fig. 1, pot #3), with slight irregularities in the contours so that the fit was not absolutely tight. The fire would have been built around and eventually over the stack, using a fairly clean-burning fuel that inevitably produced some reducing gases. Carbon dioxide from the decomposition of calcareous material in the Ungritted ware clay body may also have contributed to the reducing gases.These gases penetrated under the rim of the third pot and were trapped inside it. When the potters judged that sufficient time and temperatures had been reached, they allowed the fire to burn down, with ash and coals collecting around the base of the first pot. At the end, pot #1 would have dark fire-clouding against an oxidized background on the exterior and, if the coals and ash were heavily piled up around it, a dark gray or black base from reduction and sooting. Pot #3 would have a uniformly oxidized exterior, since it was elevated above the fuel and coals in the final cooling stage. It's interior, along with the interior of pot #1 and the whole of pot #2, would be gray, from exposure to the lightly reducing gases trapped inside. If the fit between #1 and #3 was very tight, and allowed no gases to penetrate, or if #3 cracked during firing and admitted oxygen to the interior, then the interiors of #1 and #3 and all of pot #2 would have oxidized.

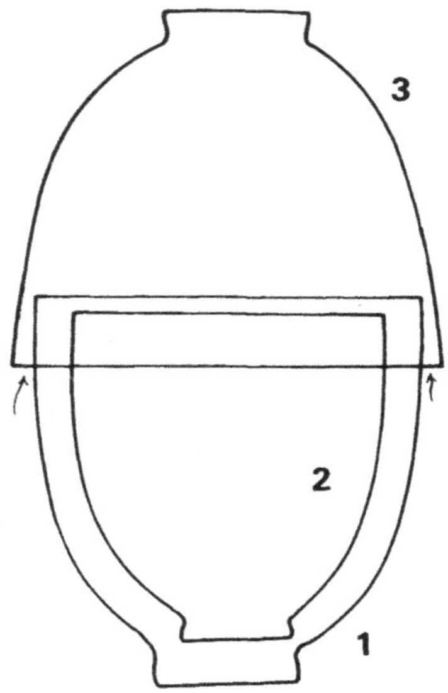

Fig. 1. Potential stacking arrangement of Ungritted ware during firing.

If taller stacks, involving more pots were used, or if pot #3 had been stacked right side up over pot #1, I would expect to find traces of firing circles (see below), and none has been observed in thousands of sherds handled. Experimental pots made with clays dug at sites that produce large quantities of this ware have very similar colors and properties and do produce firing circles when fired in stacks. Single pots could have been fired right side up or upside down to produce the most frequent, mottled, fire-clouded pieces. While other procedures may be hypothesized for production of the mottled and uniformly light-colored pieces, the uniformly pale gray sherds would have had to be covered with another vessel placed so as to trap reducing gases. The pale gray pieces were probably produced by using some pots as saggars.

Sherds from the relevant excavations were selectively discarded, so we have no reliable figures to test the 3-stack hypothesis. Limiting stacks to three vessels in height makes some sense, especially given the size and shape range in which this ware was made, since taller stacks would have been rather unstable. Impressionistically, the relative frequencies of the varieties, with the mottled being most common, the uniform light next, and pale gray the least common, appear to conform to the model. Many of the gray pots are smaller than the pots fired to lighter colors.That the potters noticed these color differences and seem to have understood their relationship to placement in the fire is suggested by several observations. First, the uniformly gray pieces have relief decoration applied before firing far more commonly than the other varieties, suggesting an intentional correlation between the two. Secondly, the rare examples with painted patterns are all fired to uniformly light colors (red on pale yellow), the firing that shows the decoration to best advantage. That the colors may also have been endowed with symbolic meaning may be further suggested by the fact that the small number of vessels in this ware that made their way south to Franchthi are all of the gray variety.

Archaeologists have long noticed and used color as a major identifier of particular wares, so we should not be surprised to find that prehistoric people also noticed this most visible characteristic of their pots, and may have assigned function, value, or other meaning according to color. While the fired colors may have been, initially, accidental consequences of the firing techniques, the existence of a number of different techniques, and differently colored results produced even at the very beginnings of the craft suggest that it probably did not take long for potters to learn some degree of control over the color of the pieces they produced. They must also have noticed early on that pieces fired inside of other pots broke much less frequently than pieces fired directly in contact with the fuel. Pots fired in direct contact with the fuel, whether or not they are enclosed in some form of heat-retaining chamber, are subject to uneven and sudden changes in temperature. In the early stages of firing--before the ware itself is uniformly heated to ca. 500° C--these sudden changes cause most of the breakage that occurs in direct fuel firings. Removing the ware from direct contact with the fuel, whether by stacking inside another pot or in the separate ware chamber of a true kiln, makes it relatively easy to warm the pieces gradually and uniformly.[2] Careful observation of the early accidental results, for potters motivated to reach for new and more stunning or reliable results, might easily have provided the inspiration and the requisite knowledge to design a true kiln, with separate chambers for pots and fuel.

MIDDLE NEOLITHIC FIRING PRACTICES

In southern Greece that step seems to have taken place in the next phase, the Middle Neolithic, sometime shortly before 5500 B.C. (calibrated). The Middle Neolithic ceramic tradition develops gradually from that of the Early Neolithic. The phase is marked by the appearance of a new ware that is a logical and technological outgrowth of earlier potters' practices and knowledge, and occurs in substantial numbers over the whole of southern Greece. The German excavator Emil Kunze named the ware "Urfirnis" or "first glaze" ware (Kunze 1931:31) because of the glassy sheen on the surface of some of the finest pieces. Recent SEM examination has confirmed that some of the Urfirnis paint reached "total vitrification," and even the clay body has often been "fired to produce at least initial vitrification" (Panakleridou et al. 1985:1, 2).

Urfirnis Ware

The Middle Neolithic Urfirnis is an impressive ware produced by skilled potters, some of whom knew how to manipulate their materials for maximum effect, both before and during firing. The clay body is reasonably rich in iron oxides and was gritted with very fine particles of calcium carbonates, under 1 mm in maximum dimension. To judge from the shapes produced, it must have been a plastic body with good green strength and shrinkage well controlled by the plentiful, if very fine grained nonplastics. All vessels made in the ware were painted, either in a monochrome coating or in patterns, with an iron oxide-rich slip (Panakleridou et al. 1985:2). The slip generally had low viscosity, for it allowed application in long fluid strokes, and was sometimes applied so thinly that what look like individual brush hairs are visible in each stroke.

The earlier pieces of Urfirnis rarely fired to produce a glossy paint. The fired colors run the gamut of low-fired iron oxide-rich clays: from black to red and everything in between for the paint. The fabric colors are less intense and range from gray to pinkish tan. Cores are generally uniform--usually light colored, but gray if the surfaces are gray or black. In the earlier pieces, fire clouds are common and lime pops are rare, while in the later, more frequently sintered or vitrified pieces, clouds are less consistently present and lime popping, sometimes extreme, is fairly common on vessel exteriors. In the vitrified pieces, colors may include tints of blue, green, and lavender.

Some Urfirnis sherds have clear firing circles (Fig.2). These are clearly defined circular areas, concentric with the base or bottom, and more rarely, with

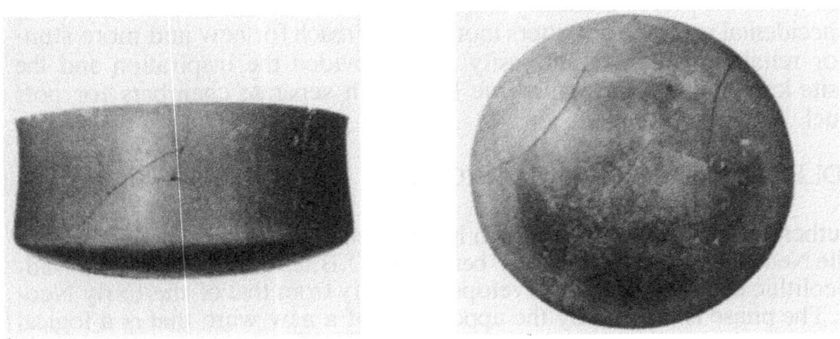

Fig. 2. The paint on a Middle Neolithic Urfirnis ware bowl from Franchthi has fired red on the exterior walls (left), with a black firing circle on the exterior bottom (right; rim diam.0.16 m).

the rim, in which the color beyond the circle contrasts sharply with the color within. Firing circles result from stacking vessels during firing in a way that exposes the portion of one vessel that is covered by the rim or base of another to a firing atmosphere different than that around the uncovered vessel parts. The very clear circles on the Urfirnis sherds indicate that pots were fired in stacks, similar to those used by the Early Neolithic potters who produced

Ungritted ware, but probably involving more vessels. A stack allows the potter to fit more pots into a smaller space and elevates some so that they might better catch the rising heat of the fire. At first glance their use seems to represent a more efficient use of fuel and perhaps an attempt to reduce firing breakage. But pots inside of other pots mean that the potter has to maintain maximum firing temperatures for longer to insure that pots on the interior of the stack receive sufficient heat, without exposing those on the outside of the stack to excessive heat. Potters probably learned to judge the timing by experience. Underfired pieces might have been refired. Early attempts to compensate for underfiring may be what prompted the higher firing temperatures and/or longer firing times that accidentally produced the occasional early pieces with glassy paint. Those early accidents provided the opportunity to observe the glassy effect and could have offered inspiration and example for the procedures used by the later Urfirnis potters who managed to achieve the glassy result with considerable consistency.[3]

Fig. 3. Middle Neolithic Patterned Urfirnis ware vessel from Franchthi in which the iron oxide-rich paint has fired black against the oxidized background clay (rim diam. 0.24 m).

Another probably accidental outcome of the early firings in stacks, and also diagnosed and reproduced intentionally by the later Urfirnis potters, was the occasional early piece in which the iron oxide-rich paint reduced to black, while the clay fabric oxidized to a light tan. This combination provides a much more dramatic effect, especially on patterned pieces, than the more usual red color against a tan background (Fig. 3). In the later phases of Urfirnis production it is the more common combination and, since it was more difficult to produce than an oxidized red paint, was presumably the preferred, intentional outcome.

The black-on-light combination with iron oxide pigments must have been produced by a process similar to the three-stage firing described by Joseph Noble for the better-known classical Greek black and red figure pottery (Noble 1988:79-80, 155-156). The clay used to make the pots and that used for the slip contain the same iron oxides. In the initial, oxidizing, stage of firing, both would have turned light colors. When the temperatures had reached ca. 800° C, the second stage was initiated by changing the atmosphere to reducing. Noble suggests that the classical Greek potters accomplished this stage by stoking with quantities of wet sawdust or green wood, and cutting off the oxygen supply by closing up all the air vents to their kilns. He says it is important that this reducing phase be maintained for at least half an hour, during which time the temperature should be gradually raised to 945° C, and then slowly lowered to ca. 900° C (Noble 1988:156).

During this reduction stage of the firing, the entire vessel turns black, but the finer particles of the slip also begin to sinter and become encased in a glassy or quartz-like layer (Noble 1988:80). Next comes the final stage, in which oxygen is readmitted to the firing chamber while the temperature is still around 900° C, by opening the air vents and slowly allowing the kiln and the ware to cool. During this final oxidizing stage the porous clay body reoxidizes and turns light again, but the sintered paint, encased in its quartz-like layer, does not permit the re-entry of oxygen at the available temperatures, and remains black.

Urfirnis Kilns

The classical Greek potters left us, conveniently, drawings of their stationary updraft kilns, and of potters closing the vents, on a number of black figure ceramic plaques from Ancient Corinth (Noble 1988:151, especially Figs. 232-234). Unfortunately, the Neolithic potters provided no such illustrations, nor have we yet identified any remains of whatever structures they might have used some 5000 years earlier than their classical counterparts. The Neolithic pots with glassy, black-painted patterns on a light background, however, testify to their knowledge of a similar process. They may have built their kilns at some distance from their settlements, perhaps to be close to fuel supplies, and we have simply not found them yet. Perhaps their kilns involved temporary or ad hoc constructions that have not survived (see below). Another possibility occurred to me when I came across the following passage in DeBoer and Lathrap's ethnoarchaeological account of the Shipibo-Conibo potters in the Amazon.

> Pottery is fired in several different ways. Food bowls, beer mugs, *shrania*, and small-sized jars are fired singly in a *mapú ëite*. The unfired vessel is inverted inside the bottomless *mapú*

ëite and covered with wood ashes (DeBoer and Lathrap 1979:120; emphasis added).

The authors provided no illustration of a *mapú ëite,* but their passing mention of a bottomless pot in which other pots are fired suggested an explanation for two puzzling questions I had about the Greek Urfirnis. The first has to do with sherds from the variety known as Coarse Urfirnis, and the second, with the procedures for their three-stage firing.

The Coarse Urfirnis sherds found at all southern Greek Neolithic sites are somewhat thicker than the sherds from other Urfirnis varieties, and regularly contain both more and larger nonplastic inclusions. The sherds all come from large vessels, with rim diameters ranging from ca. 0.50-0.90 m. The rim sherds are of two types: one (Fig. 4a) is a thick, folded-over splaying rim, with diameters ranging from 0.70-0.90 m. The other (Fig. 4b), has a more or

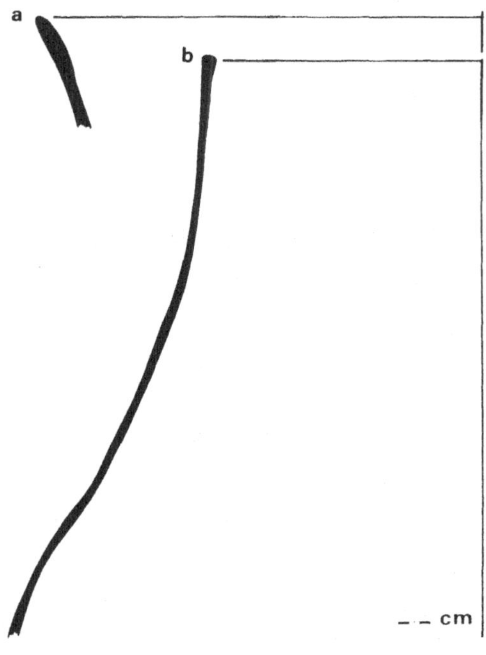

Fig. 4. Rim profiles of Middle Neolithic Coarse Urfirnis ware vessels from Franchthi.

less vertical neck, squared rim tip and rim diameters of 0.45-0.60 m. Several long profiles show the neck curving out to a wide belly, starting ca. 30-40 cm below the rim, and many examples include double strap handles ca. 20 cm below the rim. No complete profiles of these large vessels have ever been restored. The sherds in this variety show very little curvature, other than the concave-convex curve of the area below the vertical neck. Indeed, I have never seen, in the hundreds of thousands of sherds of this ware that I have sorted, a single Coarse Urfirnis sherd with either a sharp angle or sufficient curvature to suggest that it came from the bottom of even an extremely large vessel.

The Shipibo-Conibo's bottomless *mapú ëite* suggested a solution: a bottomless Coarse Urfirnis vessel, with the vertical rim as the upper portion and the splayed rim providing the lower, resting surface (Fig. 5). On paper the dimensions work out quite nicely and produce a bell-shaped vessel that would

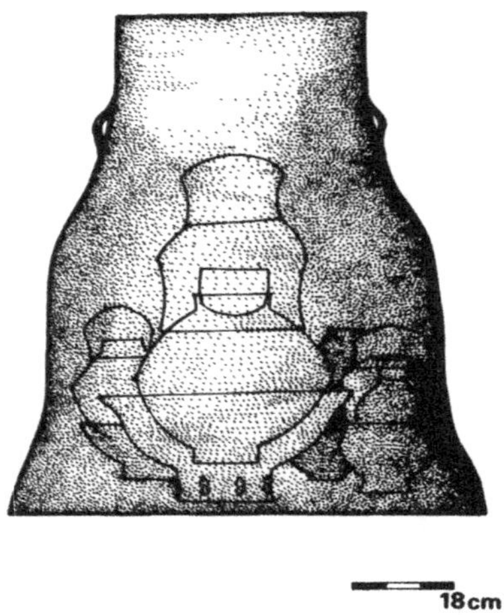

Fig. 5. Hypothetical reconstruction of a Middle Neolithic portable kiln based on Coarse Urfirnis rims from Franchthi.

have been roughly 1.00 m tall. Again on paper, this new vessel fits neatly over the largest Urfirnis vessels we have, and equally well over a substantial stack of vessels. The handles are, in fact, neatly placed so as to provide good balance of the upper part while grasping the vessel from the lower rim for lifting or lowering. While I can think of, and have received suggestions for, a number of uses for a large bottomless cylinder, the most attractive one is that it was a portable, reusable part of the kiln the potters must have used to fire their other pots.

Experimental Test of Potential Urfirnis Kilns

Filling this bottomless pot with ashes, however, would hardly have produced the necessary temperatures, much less the other characteristics noted for the Urfirnis sherds. Several years ago I presented this combination of evidence as a challenge to an experimental ceramics class that, fortunately, included 5-6 graduate students in ceramics. Unlike the beginning potters among the archaeologists, the more experienced potters were undaunted by the prospect of building such a large vessel and eager to take on the challenge of making it work as a kiln. When we finally got a warm, dry day suitable for firing, they presented me with three plausible solutions, none using an exact reproduction of the Neolithic pieces, but all reasonably close, effective, and surprisingly efficient.

The simplest version (Fig. 6) used the smallest dimensions of the Coarse Urfirnis pieces, and fired a stack of only five pots. This design supported the bottomless cover and the unfired pots on a round pisé (clay mixed with straw) shelf, pierced with multiple vent holes and in turn supported on three pisé bricks. The pots were stacked on the shelf, the cover added, and the joins sealed with additional moist clay. Chimney sections were formed and added as needed during the firing to control and improve the draw. In fact, the open stoking area around the tripod supports at the base proved inefficient and in the early stages of firing was sealed up on two sides with additional bricks and sherds from previous firings. That minor modification greatly improved the draw.

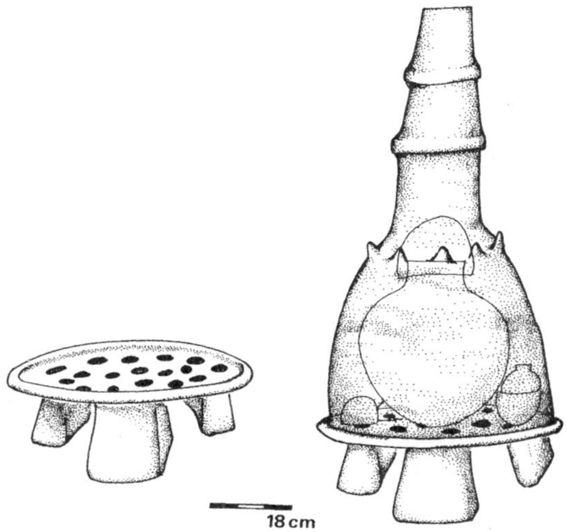

Fig. 6. Experimental bottle kiln supported on a tripod base.

The second design, dubbed the "feminist kiln" for its elaborate symbolic decoration,[4] had a somewhat more elaborate and effective system for containing the fire and raising the pots a bit higher above the heat source. At ground level were two concentric pisé cylinders (Fig. 7): a small one to provide support for the ware shelf at the center, and a larger one outside it, also to support the shelf and, additionally, to contain the fire. The design included

Fig. 7. The experimental "feminist kiln," supported on a base of concentric cylinders.

four arches cut from the exterior cylinder, to provide stoking channels.The cut-out portions were replaced at the end of the firing to close these vent holes and contain the heat. On top of the pair of cylinders, the designers placed a series of wedge-shaped, pierced pisé shelves. 30-35 unfired pots were stacked on these. The bottomless cover was built in two sections--largely because the team procrastinated on the actual execution of their design and didn't allow themselves enough time to dry and scrape the walls to arrive at a closer approximation of the Neolithic model. Both sections of the cover were added over the stacked pots, a few additional small pots were fitted into unfilled spaces, pisé chimney sections were added, and all the joints were sealed with moist clay. Because this kiln was still damp when the firing began, it developed numerous cracks during the firing, but these were easily sealed with additional clay.

The third design used simple pisé bars to provide a grill-like shelf for the unfired pots, and supported them on the rim of a dug hole (Fig. 8). This hole served as the fire chamber, fed through a tunnel that connected it to an adjacent sloping hole. Pots were stacked on the bars, and the stack was covered with the bottomless cover, again built in sections, with added pisé chimney pieces. In this kiln, a concerted effort was made to match the diameter of the entrance flue to that of the exit, or chimney flue. The kiln was also taller, relative to its diameter, than the other two versions, and it proved to have the most efficient draw of the three. Unfortunately, it was built of untempered commercial clay (the other teams had dug their clay at the firing site and tempered it liberally with straw), was still quite damp when the firing began, and exploded quite dramatically well before the firing was complete, at approximately 500° C. The pots inside, however, suffered little damage, and were fired in a subsequent firing.

Fig. 8. Experimental bottle kiln supported on the rim of a dug hole.

We had only one pyrometer available for this firing and used it in the feminist kiln. Fueled with assorted sticks and twigs under 5 cm in diameter, of local hardwoods, it took just two hours to reach 850° C in the center of the stack of pots.[5] We lost not a single vessel from this one-stage, oxidizing firing. Pieces that had been placed at the bottom of the stack, directly over the holes in the shelving, showed some fire-clouding. Pieces from the upper parts of the stack were uniformly oxidized. No firing circles were, however, evident on our pieces.

Without additional pyrometers, and with no way to see the pots under their bottomless covers, there was much speculation about how we could judge when the pieces were fired. The students were most pleased with what they called the "Orton Dog" measure. Comparing results from the kiln with the pyrometer and those without, they came up with a series of temperature indicators such as "at 800° C a hot dog held directly over the chimney blisters in under 5 seconds." More seriously, in the early stages of the firing, a hand passed quickly over the chimney provides a good sense of heat and moisture in the chamber. As the temperature increases, small twigs held over the chimney take decreasing amounts of time to ignite. At 800-850° C the twigs turned almost instantly to ash. With more experience, I'm sure we would have noticed a number of other useful indicators of temperature.[6]

As luck would have it, our firing came to a somewhat premature end because of the unpredicted (but entirely typical) arrival of a midwestern spring thunderstorm. Those firing the simple kiln had planned to attempt a reduction firing. As the storm approached rapidly, and with their kiln at only an estimated 800-850° C, instead of the hoped-for 900° C, they poured damp wood chips into the chimney, filling the ware chamber, and then sealed the chimney. They stuffed more wood chips into the fuel box, sealed it with clods of earth, plugged any cracks that were seeping smoke with handfuls of wet clay, and left the smoking kiln to the rains and the night.

We returned the next morning to unload the kilns and found that during the night, and in spite of the heavy rains, the temperature inside the sealed, sawdust packed kiln must have remained high for some time, and cracks had opened up to re-admit oxygen to the ware chamber.The wood chips had caught fire and burned to ash. When we unloaded, only those pieces at the very bottom of the stack, pieces covered by the ash from the wood chips, were black. The upper pieces had reoxidized in what was, essentially, the third stage of an unintended three-stage firing. The accident suggests how the Neolithic potters, aiming for black pots, might have chanced on the procedure that produced the more dramatic result.

We did not then try to execute, and have not in subsequent attempts succeeded in executing, a three-stage firing that produces black iron oxide paint on a light ground. Our major difficulty so far is in maintaining, much less raising the temperature during the reduction stage, although even when we finally succeed at that, we may find that our clay slip mixture still does not have the specific and I suspect rather special properties of the Greek version used by Neolithic and later Greek potters (Noble 1988:85).

I should also note that the rainstorm contributed to another archaeologically interesting outcome. Excavations at Neolithic sites have identified pits, sometimes in pairs and with indications of burning. These could have been part of a two-holed dug fire chamber similar to the third experimental example. We find broken pots that could have been reused as chimney segments in the manner illustrated on the drawings of classical Greek kilns (for example, see Noble 1988:151, Fig. 233). But aside from these rather vague indications and a few pieces of burned daub very similar to what we produced from the plugging of seams and cracks during our firings, we have no evidence from Greek Neolithic sites for the constructions that the students used to elevate their bottomless covers above the fire and to support the unfired pots. In the firings described above, and in subsequent experimental firings, we have found that well-built ware chambers, especially if thoroughly dried before their first firing, survive and can be reused. The lower, supporting segments, however, tend to crack and crumble. The rainstorm had already started to slake our pisé pieces before we began to unload, and by the time we had dismantled and unloaded completely, these elements had been reduced to shapeless crumbs. We have also found that these crumbs and any other parts of our kilns that are too badly damaged to reuse in another firing are easy to pulverize and provide an excellent source of grog--a practice that in turn provides a lovely rationale for the absence of any evidence for these pieces in the archaeological record.[7]

We have refired several of our kilns as many as three or four times, each time rebuilding the lower and chimney sections. In almost every case, the reason we have stopped using a kiln was not that it ceased to be functional, but that the empty open space where we were permitted to build and fire was co-opted for some construction project, and our kilns have been bulldozed into oblivion. One wonders if prehistoric potters were also victims of "progress" and settlement expansion and had to continually relocate their firing areas.

We may yet find and excavate the kilns that the Middle Neolithic potters used to produce their three-stage controlled firings, and they may prove to be nothing like the "bottle-kilns" my students produced. But the sherds the Greek Neolithic people left us speak pretty clearly to the fact that by the 6th millennium B.C. the southern Greek potters had developed rather sophisticated firing techniques. These included the use of some kind of kiln that removed most of their pots from direct contact with the fuel, that permitted them to achieve and maintain relatively high temperatures and to control the atmosphere and, hence, the colors of their pots. The potters were producing on a relatively small scale, and few of their pieces were, apparently, destined for exchange. These are among the factors that have led me to suggest that their pots, and perhaps even the production process, were geared primarily for use in the symbolic arena of Neolithic life (Vitelli 1995:61-62). While their Urfirnis ware may be an unusually sophisticated product, motivated by the particular circumstances in which the potters worked, and made with unusually color sensitive materials compared to most other early prehistoric potters, I hope the examples cited here might encourage other archaeologists to consider the potential evidence for prehistoric and other early firing practices recorded in our substantial collections of sherds.

NOTES

[1] The following discussion is based on the author's observation of thousands--and for the Middle Neolithic, hundreds of thousands--of sherds in each ware, along with over twenty years of experimental pottery making. By law, the excavated sherds may not leave the storerooms in Greece without special permission. With such permission, characterization and provenience studies have been conducted on a very limited sample of sherds from the Franchthi excavations. For results and discussion of these studies see Jones 1986:386-402; Cullen 1985:233-235, 399-415; Vitelli:1993:13-19.

[2] In our experimental firings, we have learned to be pleased with 75% recovery of intact pieces from direct fuel firings. From kiln firings we expect 90-100%.

[3] I have not yet had the opportunity to make close observation of coloration on large numbers of Urfirnis ware sherds from other sites to explore, for example, the consistency in firing among sites (Frankel 1994:205).

[4] And because the men in the class had teamed up to build a long "macho" climbing kiln on the nearby hillside. Each team independently decided to decorate its kilns, all claiming that it was "to bring good luck." Interestingly, several of the Coarse Urfirnis pieces have painted decoration, and all have, on the interior surfaces, "dapples," or splotches and drips of paint that look as though flicked from the fingers in an expressive gesture--perhaps also for good luck?

[5] When the feminist kiln had begun to cool down, we moved the pyrometer to the climbing kiln, inserting the wire through a peephole at about the midpoint of the ware chamber. After six hours of continuous stoking, that kiln had reached only 600° C, and it took a full 12 hours to reach 900° C.

[6] We have also found that strips of venison or other raw meat suspended on skewers over the chimney can be smoked during the warming stages, or cooked when the heat increases, and are extremely tasty.

[7] Analyses done for a colleague, Tracey Cullen, of a sample of Urfirnis ware sherds mention the possibility of grog inclusions in a few sherds (Pomoni-Papaioannou 1982, Cullen 1985:229-233).

REFERENCES

Cullen, Tracey.
1985. *A Measure of Interaction among Neolithic Communities: Design Elements of Greek Urfirnis Pottery*. Unpublished Ph.D. dissertation, Program in Classical Archaeology, Indiana University, Bloomington.

DeBoer, W.R. and D.W. Lathrap.
1979. The Making and Breaking of Shipibo-Conibo Ceramics. In *Ethnoarchaeology. Implications for Archaeology*, edited by Carol Kramer, pp. 102-138. Columbia University Press, New York.

Dowman, E.A.
1970. *Conservation in Field Archaeology*. Methuen, London.

Frankel, David.
1994. Color Variation on Prehistoric Cypriot Red Polished Pottery. *Journal of Field Archaeology* 21:205-219.

Jones, R. E.
1986. *Greek and Cypriot Pottery. A Review of Scientific Studies.*. The British School at Athens, Fitch Laboratory Occasional Paper 1. The British School at Athens Publications Department, Athens, Greece.

Kunze, Emil.
1931. *Orchomenos, II. Die neolithische Keramik.*. Abhandlung der Bayerischen Akademie der Wissenschaften, Munich.

Noble, Joseph Veach.
1988. *The Techniques of Painted Attic Pottery*, revised edition. Thames and Hudson, New York.

Panakleridou, D., A. A. Katsanos, and Y. Maniatis.
1985. Technological Examination of Urfirnis Pottery. N.R.C. "Demokritos," Athens. Ms. on file, Program in Classical Archaeology, Indiana University, Bloomington.

Pomoni-Papaioannou, F.
1982. Results of the Petrographic Analyses of Neolithic Sherds from the Sites of Asea, Ayioryitika, Corinth, Lerna, and the Franchthi Cave. Institute of Geological and Mining Research, Athens. Ms. on file, Program in Classical Archaeology, Indiana University, Bloomington.

Rice, Prudence M.
1987. *Pottery Analysis. A Sourcebook..* The University of Chicago Press, Chicago.

Rye, Owen S.
1981. *Pottery Technology: Principles and Reconstruction..* Manuals on Archaeology 4, Taraxacum, Inc., Washington, D.C.

Shepard , Anna O.
1968. *Ceramics for the Archaeologist..* Carnegie Institution of Washington, Publication 609, Washington, D.C.

Vitelli, Karen D.
1993. *Franchthi Neolithic Pottery*, Vol. 1. *Classification and Ceramic Phases 1 and 2..* Excavations at Franchthi Cave, Greece, Fascicle 8. Indiana University Press, Bloomington and Indianapolis.

Vitelli, Karen D.
1995. Pots, Potters, and the Shaping of Greek Neolithic Society, in *The Emergence of Pottery. Technology and Innovation in Ancient Societies,* edited by William K. Barnett and John W. Hoopes, pp. 55-63. Smithsonian Institution Press, Washington, D.C.

POTTERY FIRING STRUCTURES (KILNS) OF THE INDUS CIVILIZATION DURING THE THIRD MILLENNIUM B.C.

Heather M.-L. Miller, Anthropology, Univ. of Wisconsin-Madison,
1180 Observatory Drive, Room 5240; Madison, WI 53706-1397
heatherm@macc.wisc.edu

ABSTRACT

The Indus Valley ("Harappan") craftspeople developed a variety of ways to fire clay objects. This paper illustrates the technology of pottery and terracotta object firing in the greater Indus region during the third millennium B.C., drawing on scattered and unpublished evidence for a range of firing structures. "Open-air", single-chamber ("oven" and "pit"), and double-chamber updraft firing structures are discussed, with examples from the sites of Mehrgarh, Nausharo, Harappa, and Mohenjo-daro. I particularly stress the contemporaneous use of different types of firing structures and the continuum nature of firing structure "types".

INTRODUCTION

The Indus civilization was located in what is now Pakistan and northwest India during the third millennium B.C. Although it covered a far greater area than contemporaneous civilizations in Mesopotamia and Egypt, it has been the focus of much less research. It is chiefly known for its planned cities, particularly their well-developed water and waste-disposal systems, and for its highly talented craftspeople, who were particularly creative with various pyrotechnologies.

But Sophie Méry (1994) rightly points out that much of the published research on Indus craft industries has focused on prestige and luxury ornaments, such as long-barreled carnelian beads, steatite stamp seals, glazed steatite beads, and stoneware bangles. In contrast, this paper will focus on the production of terracotta pottery, rather than on "prestige" ceramic materials produced by Indus artisans, such as stoneware, siliceous faience, and talcose paste and faience. (The Indus peoples did not make glazed clay objects, but did glaze talcose and siliceous objects.) Pottery has always been a topic of interest in Indus research, but primarily for chronological reasons. However, the past decade or so has seen a great increase in the interest in pottery production sites, particularly with the recent spectacular finds of potters' workshops and a range of firing structures. These new data have allowed me to interpret firing structures from early, incompletely published excavations, and to begin to characterize the variety of Indus firing structures used for pottery and/or other terracotta objects.

This paper is not a gazetteer of all known firing structures, but rather examines the range of firing structures and technologies used by Indus. (The relation between firing technologies and the economics—and politics—of pottery production will not be discussed in this paper, due to lack of space.) I have chosen only the best described and/or most common types of firing structures, but have tried to portray the variety of types of structures found. Most of my evidence comes from three groups of sites: (1) Mehrgarh, Lal Shah, and Nausharo in the Kachi Plain of southern Baluchistan; (2) Harappa, on the Ravi River in the flood plains of the Punjab; and (3) Mohenjo-daro, in the southern flood plains of the Indus River itself, in Sindh. The mounded urban sites of Harappa and Mohenjo-daro are the two best-known of the half-dozen largest Indus centers and the only ones to have been excavated, while the sites in the Kachi plain have revolutionized our knowledge of the earlier periods of this region. But these are by no means the only places where possible pottery firing structures have been found, but they are by far the best published, and firing structures found at other important sites are noted wherever possible.

I focus on the third millennium B.C., preceding and during the florescence of the urban Indus ("Harappan") civilization. As the chronological terms for the Indus region vary widely, and are still being extensively revised with new excavations, I will provide broad dates as currently known, but will also include the period/phase assignments. The most important point about chronology for this paper, however, is that there is <u>not</u> an evolutionary progression of firing structure types, as far as I can tell at this point. While increasingly more complex firing structures do appear over time, variants of the older structures apparently continue to be used.

Classifications are made for different reasons; my rough classification of firing structures in this paper focuses most on the ways in which structures dealt with the flow of air or heat (draft). It matches most closely Rye's (1981:96-100) classification of *open* vs. *kiln* firing, with pit structures and ovens as variants of open structures, and "kiln" restricted to structures with separate chambers for fuel and for products. Sinopoli's (1991:31-33) classification of *hearth (open-air)*, *pit*, *oven* and *kiln* structures is also similar. Rice's (1987: 153-163) classification of *nonkiln* vs. *kiln*, where "kiln" includes pit, bank, updraft and downdraft structures, is quite different. In this paper, I will discuss three categories of firing structures: *"open-air"*, *single-chamber*, and *double-chamber updraft kilns*. There are a number of other types of firing structures, including short "cylindrical kilns" from Mohenjo-daro and Harappa (Mackay 1938:49-50; Vats 1940:470), and "jar kilns" from Harappa (Dales & Kenoyer 1991:230, 1992:60; Wright 1991:78; Vats 1940:470), which will not be described in this paper.

To date, no true "bonfire" firing structures have been recognized for the third millennium B.C. in the Indus. All of our *"open-air" firing structures* involve some type of fairly substantial covering in addition to fuel, including potsherds, earth, and/or mud plaster. My category of *single-chamber firing structures* comprises slightly more permanent structures, including both "pit" structures and "ovens". *Double-chamber updraft firing structures/kilns* are found at a large number of sites during the Integration Era or "Mature Harappan" period, *ca.* 2600 to 2000/1900 BC. Double-chamber updraft kilns are a well-known type of firing structure, and there seem to be a number of different variants in use in the Indus. I have provided general renditions of the various types of structures (Figure 1), primarily to clarify the terminology I use (this is particularly a problem for double-chamber updraft firing structures). There are almost no drawings or plans published of the excavated structures, so these renditions should <u>not</u> be taken as exact drawings of the structures—these are based on educated guesses using the published descriptions and photographs, and some are generalized compilations of various structures excavated at different sites.

As will be apparent from the descriptions, these major types form a continuum of firing structures, of which these described types are only the "average" examples. This continuum exists even for perfectly preserved archaeological structures and/or modern ethnographic examples. With the added complication of poorly preserved archaeological structures (for example, usually only the base of a structure is preserved), assignment to a "type" can be a difficult task. It can even be counter-productive; it is much more helpful to publish even brief descriptions of excavated firing structures than simply noting that a "pit kiln" was found, especially given the number of different definitions of "pit kilns". More detailed descriptions are even more useful, such as:

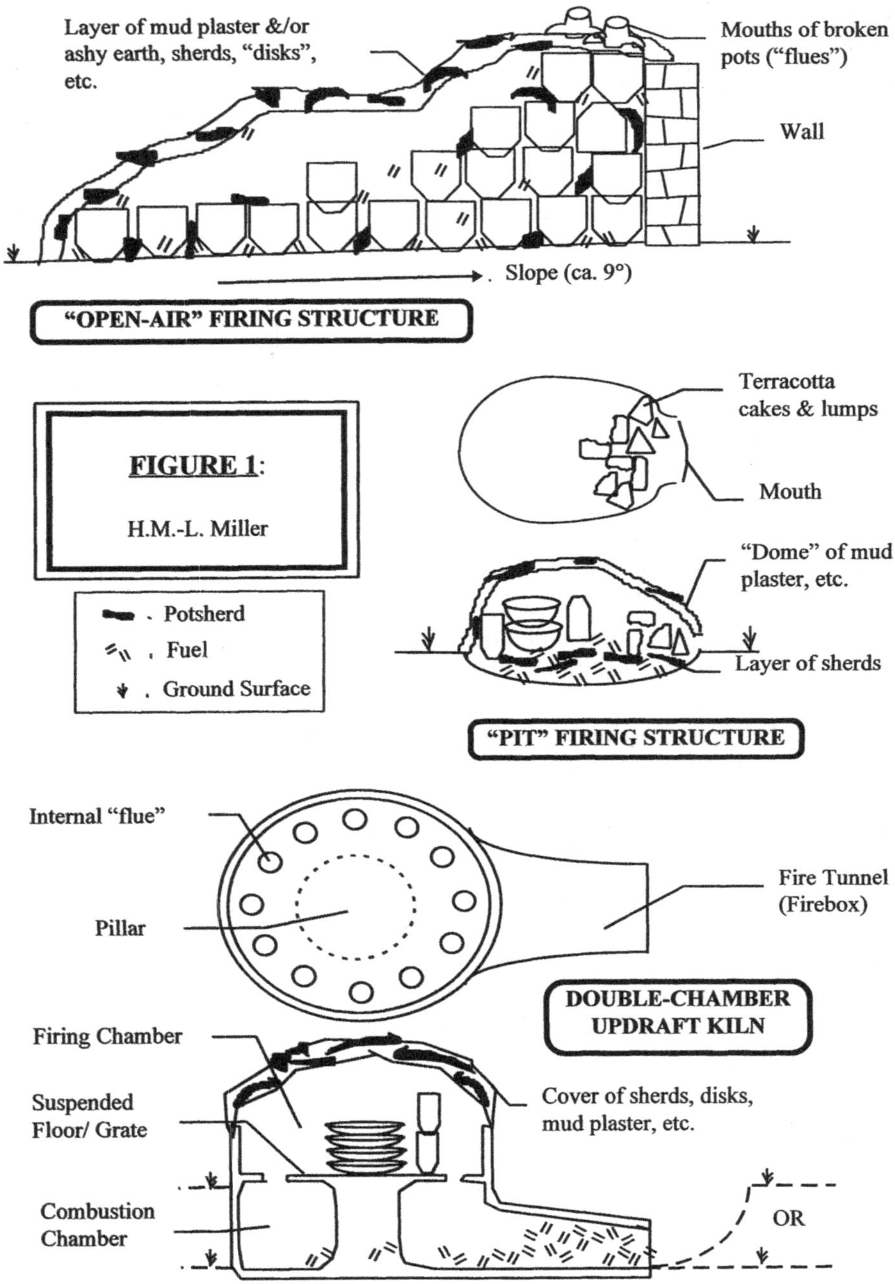
Layer of mud plaster &/or ashy earth, sherds, "disks", etc.
Mouths of broken pots ("flues")
Wall
Slope (ca. 9°)
"OPEN-AIR" FIRING STRUCTURE
FIGURE 1:
H.M.-L. Miller
Potsherd
Fuel
Ground Surface
Terracotta cakes & lumps
Mouth
"Dome" of mud plaster, etc.
Layer of sherds
"PIT" FIRING STRUCTURE
Internal "flue"
Pillar
Fire Tunnel (Firebox)
DOUBLE-CHAMBER UPDRAFT KILN
Firing Chamber
Suspended Floor/ Grate
Combustion Chamber
Cover of sherds, disks, mud plaster, etc.
OR

- Dimensions and shape of preserved structure;
- Materials used in construction (mud bricks, perhaps baked *in situ*, pre-baked bricks, stone, plaster of various types with various tempers, specially made elements such as arches or stretchers for suspended floors);
- Any associated firing "furniture" (setters, separators, saggars, suspended floor elements, etc.) and the exact location within the structure where found;
- Fuel materials found (charcoal/ash from straw, brush, twigs, large pieces of wood, dung, etc., including species used)—or at least notes about the type of material (wood, straw, dung) and size of charcoal (twigs, large wood);
- Radiocarbon dates on fuel materials, palaeomagnetic dates on clay walls;
- Temperatures reached (based on analysis of structure itself or its products);
- Location (on slopes or flats or in hollows; orientation to prevailing winds; near roads or rivers (transport); within or outside of buildings, residential areas, and/or settlements);
- Any associated object production tools (chucks/molds, burnishers, etc.) or other evidence for object production in the vicinity of the firing structure;
- Products fired (based on wasters and/or unfired objects found in or near the firing structure). For the Indus the products fired are important for dating as well as production information, because with the lack of direct radiocarbon dates for most of the structures excavated, the products fired will have to supply the chronological information to date the structure.

Another line of evidence for the technology of pottery production is analytical studies of the finished product, and several Indus researchers have made use of this approach, particularly Wright (Wright 1985, 1986, 1989a,b,c; also Jenkins 1994; Méry 1994). As this is a paper focused on the firing structures themselves, I will not discussed this approach except in passing, but it provides a wealth of information about the technology of production, including firing.

Finally, for many of the firing structures excavated in the Indus region, it is difficult to say what was fired in the structure. Especially for early projects, excavators often did not provide information about wasters in or around the firing structures. But beyond this, the overlap in the characteristics of pyrotechnological structures used for many different functions makes it difficult to determine, for instance, whether a structure was used for craft production or for cooking (C. Jarrige 1994:288; C. Jarrige *et al.* 1995:511). This has led to widely different interpretations of almost identical features. For example, a group of structures found at numerous Indus sites are all described as shallow ovate pits (ranging from 0.35 to 1 m in length), showing traces of firing, with a cylindrical or rectangular block/pillar in the center (sometimes made from a single brick coated with clay), and usually containing terracotta cakes or lumps as well as ash and charcoal fragments. These apparently identical structures are variously interpreted as ritual "fire alters" at Kalibangan (Lal 1979:77; Thapar 1973:101), as "cooking hearths/ovens" at Harappa (Meadow & Kenoyer in press), and as "fire-pits" related to domestic use (but also compared to pottery kilns) at Nausharo (C. Jarrige 1994:288).

However, the vast majority of the firing structures presented here have quite good evidence for their use in firing terracotta vessels. A variety of other terracotta objects, such as figurines, bangles, and beads, were probably fired in some or all of the firing structures described, either with pottery vessels or separately, but there are few references to finds of such objects as wasters or in firing structures. Like the vessels, other terracotta objects come in a range of qualities, so that a range of firing regimes from expedient to elaborate could have been used.

"OPEN-AIR" FIRING STRUCTURES

The "open-air" structures were found primarily at the site of Mehrgarh. The time depth of the Indus civilization was radically changed in the 1970s by the French Mission discoveries at the site of Mehrgarh in the Kachi Plain, with sequences from the seventh millennium B.C. to the third millennium B.C. at Mehrgarh and the nearby later site of Nausharo (C. Jarrige *et al.* 1995). The finds at Mehrgarh and Nausharo have played a pivotal role in research into the production of many materials, including pottery. Among the most spectacular have been finds of several potters' workshops, including several at Mehrgarh and one at Nausharo, as well as a number of separate pottery firing structures from various time periods, some within the settlements and some outside them. Finds from the workshops include unfired and fired pottery of a wide variety of types; clay scraps from the shaping of vessels and scraping of leather-hard vessels; stone, bone, and fired clay tools from both of these operations (which have different microwear patterns); fired clay chucks/molds; grinding stones, and a variety of other objects from various production stages. (See Anderson-Gerfaud *et al.* 1989; C. Jarrige *et al.* 1995; Méry 1994; Quivron 1994; Russell 1995; Santoni 1989)

The most detailed descriptions of any of the firing structures from Mehrgarh/ Nausharo are of an "open-air" firing structure, a very simple structure which nevertheless resulted in some rather sophisticated uses of heat and air flow. (Figure 1) The most spectacular example is from Mehrgarh Period VI, from the first half of the 3rd millennium B.C. (prior to 2700 B.C.), in Mehrgarh Area 1, MR.C, Square 9H. (Audouze & C. Jarrige 1979; C. Jarrige & Audouze 1979; C.

Jarrige *et al.* 1995:109,136; J.F. Jarrige 1979) The Period VI structure consisted of some 140 jars piled on a south-to-north sloping surface (sloping 9°, downhill to the north), in an oval shape 6 m long (NW-SE) by 4 m wide. The bottom of the structure consists of the remains of a mud-brick wall from the previous period on the southern end (upslope), and perhaps a paving of limestone cobbles. Above this, the excavators note that the jars seem to be set on or in a layer of clean, hard clay. The intact eastern profile of this structure revealed that the jars were piled up in as many as 4 layers, some piled one inside another, others stacked in alternate rows, and at least two jars upside down. The jars are poorly fired, some showing blistering, vitrification, or crazing. Although most were complete, they are cracked and broken in place. Numerous sherds are packed among the jars, many apparently fragments of older, broken vessels "used to stabilize the unfired jars as the layers were built up" (Audouze & C. Jarrige 1979:215). The uppermost layers of this structure consist of an ashy, powdery sediment containing "a large number of jar fragments, sherds of a finer variety of pottery, and fire-cracked limestone pebbles" (Audouze & C. Jarrige 1979:215), with increasing amounts of ashes and charcoal towards the bottom of the structure. The sherds probably functioned as coverings for the top of the firing structure. The pebbles may have functioned for heat retention (see below, "pit kiln"), or were included to take advantage of the heat of the structure rather than fired separately for lime production, or both. (See Audouze and C. Jarrige (1979:219) and C. Jarrige *et al.* (1995:184) for a description of lime production near this structure.) Around the margins of the structure, and stuck onto some of the jars or fragments of jars at the top of the structure ("especially along their mouths"), were fragments of baked clay mixed with straw (Audouze & C. Jarrige 1979:215). Most of this straw-tempered clay mixture is undoubtedly the remains of a covering plastered over the jars during firing, to retain heat within the structure. However, the straw-tempered clay around the mouths of the jars may relate to their possible use as containers, as suggested by Méry (below).

The vast majority of the jars are standardized in shape: wheel-made, with a flat base of 10-14 cm diameter, slightly rounded walls, and a wide mouth forming the maximum diameter of the vessel (around 30 cm). They are undecorated, but their color ranges widely from brick red to bottle green, both from one jar to the next and within a single jar, due to uneven firing. There are three groups of exceptions: (1) a few jars of the above type, but decorated with geometric patterns in red, black and white, (found in three different parts of the firing structure); (2) a few jars clearly larger than the rest, and (3) a dozen small jars with a ring base, carinated underside, and straight walls. (Audouze & C. Jarrige 1979:216) At least some of the painted vessels and finer vessels were found in the northern area, where the fire was probably first lit (C. Jarrige *et al.* 1995:136). Some of the jars bear different "potter's marks" (J.-F. Jarrige 1979:83).

Provisional measurements on samples of unbaked local clay and on well-baked jar fragments gave firing temperatures of 850°C in an oxidizing atmosphere. Similar measurements of overbaked fragments and underbaked fragments gave results of up to 1000°C and only 750°C, showing the irregular distribution of heat in the structure. (The method of analysis is not given, but the low temperatures were attributed to a clay composition rich in illite and limestone.) (Audouze & C. Jarrige 1979:217; C. Jarrige *et al.* 1995:136; Note that C. Jarrige *et al.* 1995:184 gives ranges of 500 to 800°C, with a "correct" firing temperature of 600°C.)

This firing structure was interpreted with the aid of modern structures of startling similarity created by professional potters in the Kachi plain to fire large numbers of vessels. The modern potters set as many as 500 to 1000 vessels on a sloped surface covered with a layer of straw and other agricultural by-products (C. Jarrige & Audouze 1979:91), which is often terminated by a wall at the upper end. The vessels are then themselves covered by a layer of straw and other agricultural waste, then by potsherds, and finally covered over by ashy earth saved from previous firings and/or by a layer of clay plaster. The fire is lit from the lower end ("in a sort of open oven built at the lower end"), and the natural uphill draft passing between the jars is encouraged by using the mouths of broken jars to form ventilation chimneys at the upper end of the structure, rather like the principle behind East Asian bank kilns. The firing lasts 24 hours, with no addition of fuel, followed by a week of cooling. The firings are done during the summer months (April-Sept.) after the winter rains are over. (Audouze & C. Jarrige 1979:219-220; C. Jarrige & Audouze 1979:91; C. Jarrige *et al.* 1995:109) The advantages of the technique are that a huge number of vessels can be fired with a limited amount of fuel, and that there is no need to maintain a permanent firing structure. The disadvantage is that there is a fairly high percentage of failures during rain or winds. (J.-F. Jarrige 1979) Presumably the archaeological example represents such a firing failure, perhaps caused by a wind or storm leading to uneven heat and flare-ups of overly high temperatures within the structure. (C. Jarrige *et al.* 1995:136)

These modern structures seem to match the archaeological structure in almost all details, down to the wall at the upper end of the structure. Allowing for the gaps in the distribution of vessels where well-fired jars were apparently removed, the original archaeological structure must have contained some 200 jars (Audouze &

C. Jarrige 1979). At least in some cases, the jars may have been used as firing containers for smaller vessels, as well as being fired themselves. (Méry 1994: 479-480,481 ftnt. 5)

A number of other "open-air" structures are reported from Mehrgarh, dating to both "earlier and later levels" than the Mehrgarh Period VI example described above (Audouze & C. Jarrige 1979:219). Most were apparently quite ephemeral, due to the removal of the (successfully) fired vessels. An earlier example is dated to *ca.* 3300 B.C. (in Mehrgarh Period V), from the same location as the Period VI structure described above (Mehrgarh Area 1, MR.C, Locus CCXXXII), and associated with two rooms containing large amounts of a wide variety of pottery types. This Period V structure is slightly different, as its base was a shallow depression rather than a flat slope, but it also functioned to fire large numbers of vessels. The Period V vessels from this structure were almost identical to the Period VI structure jars in shape, but smaller (20 cm max. diameter) and made either of heavily-tempered coarse clay or of wet ware. (C. Jarrige *et al.* 1995: 321,344,374,425; J.-F. Jarrige 1995:78)

A later example from the nearby, later site of Nausharo dates to 2500-2400 B.C. (the Period II levels of sector NS.K, in the southern mound). In a room containing six or seven layers of refuse from a variety of forming and firing stages of pottery production, "glazed coating fragments" were found "in a primary position, on the facing of the western and southern walls" (Méry 1994:474). Based on this evidence, as well as finds of 40 cm of "white compacted ashy microlevels containing few potsherds and firing wasters" extending to the south of this room, the excavators suggest that Room 2 was used (or more likely re-used) for pottery firing in an "open-air" type of structure (Méry 1994:474). Méry also notes that, as for earlier periods at Mehrgarh, this would be just one type of firing structure among other contemporaneous types, as at least one double-chamber updraft kiln dating to Period II has been found within 100 meters of NS.K area.

Another important group of firing-related objects from a different layer in Room 2 are fragments of flat, rounded clay "disks" made of a mixture of straw and clay (25-30 cm in diameter, 3-4 cm thick). More of these objects have been found in a Nausharo Period III potters' dump (*ca.* 2400 to 2100 B.C.). (Méry 1994:479) Méry suggests that, based on modern Baluchi firings, these may have been used to cover the "open-air" firing structures. Similar objects at Harappa (although somewhat thinner) have been interpreted in similar ways (J.M. Kenoyer, personal communication), and could be used to cover/roof any of the firing structure types discussed in this paper.

In sum, these "open-air" structures are a clever use of simple physical principles (a slope to increase the draft) and low-cost materials (including the use of agricultural wastes) to achieve maximum temperatures and minimal fuel use. So why was this technique not used elsewhere in the greater Indus region? It probably was, and the ephemeral nature of a successful firing has precluded its recognition. For example, Mackay (1938:109) found at Mohenjo-daro "a mass of debris several feet high" that was "evidently plastered over, for one or two patches have been preserved by the fire that also baked part of the plaster on the courtyard walls". (See also Mackay 1938:49,121 for other finds at Mohenjo-daro of burnt mud-plaster found in areas with burnt walls.) This is much like Méry's description of the remains of an "open-air" structure at Nausharo (above). However, this is not to deny the possibility of regional variation in firing techniques; perhaps at sites in other parts of the greater Indus region, pit structures were used instead of "open-air" structures. Indeed, such regional variations are quite likely, given the vast size of the Indus civilization, but at present the data is not sufficient to even speculate on regional differences.

SINGLE-CHAMBER FIRING STRUCTURES

As noted above, we are dealing with a continuum of structures; in fact, Rye (1981) refers to pit and other types of single-chamber firing as a variants of "open firing". The Mehrgarh Period V "open-air" firing structure described above might be classified as a "pit firing", given the presence of a shallow depression. Another type of single-chamber firing structure, from Mohenjo-daro, might be better classified with the open-air structures above, since it employed the use of a draft created by a sloped floor and "flues" at its rear.

Sub-Triangular Single-Chamber Structure from Mohenjo-daro

Quite different from the Mehrgarh and Harappa examples below, one of the best described single-chamber firing structures from an Indus site was found at Mohenjo-daro (south of DK-B,C area) by the IsMEO surface surveys (Pracchia *et al.* 1985; Pracchia 1987). Unfortunately, due to its proximity to the surface, it can only be dated vaguely to the "Mature Harappan" (*ca.* 2600 to 2000/1900 B.C.). However, due to careful observation and reporting, a great deal of information is available about its operation in spite of its poor preservation. Only the very base of this sub-triangular structure was found, measuring 1.40 by 1.50 m, with walls delimited by a single row of almost complete re-utilized bricks. It was possibly surrounded by a filling, perhaps indicating that it was originally semi-subterranean, but too little of the structure remains to be sure. Pracchia compares its shape to Vats' "pear-shaped kilns" from Harappa (see below). (Pracchia 1987: 153; Pracchia *et. al* 1985:219) The structure contained drops of vitrified clay from melted mud plaster, a layer of ashes, and a group of vessels in front of the stoke-hole/fire tunnel which were probably from later dumping after the structure's abandonment (Pracchia *et al.* 1985:219). There is no sign of a pillar, nor of any vitrified material from a suspended floor (Pracchia 1987:153). There are two unique aspects of this structure which support the notion of a single-chamber firing structure with a draft. One is the presence of two blocks of clay melted into place on the back corners of the structure, which appear to indicate the existence of two exterior flues or chimneys at these points (Pracchia *et al.* 1985: 215; Pracchia 1987:152,Fig. 5). The second aspect is the inclined floor of the structure, sloping upwards from the stoke-hole/fire tunnel to the back of the structure (Pracchia 1987:153,Fig. 6), perhaps to create a draft as in the "open-air" firing structures from Mehrgarh/Nausharo. This structure is thus an excellent example of the continuum nature of firing structure types.

"Pit Kiln" from Harappa

It might seem curious that only a single example of a "pit kiln" has been published for the Indus, from the recent excavations at Harappa (Dales & Kenoyer 1991, 1992). (Figure 1) This is particularly odd given that "pit kilns" are among the most common type of firing structure used in South Asia today. But on the one hand, some of the "ovens" from Mehrgarh (below), might well be classified as "pit kilns". On the other hand, many of the modern "pit kilns" in India and Pakistan would in fact be closer to either the "open-air" firing structures described above, or the single-chamber ("oven") firing structures described below, as they do not employ a scooped-out pit, but are built onto a surface. It may also be a case of these structures being overlooked, or mistaken for cooking structures, as noted in the Introduction.

The Harappa pit structure was found on the northwest corner of Mound E, one of three firing structures found in an area of some 70 square meters (Dales & Kenoyer 1991, 1992; Wright 1991). It was in deposits dating to Harappa Period 3, and is radiocarbon dated to about 2300 B.C. (calibrated) (Dales & Kenoyer 1991:43,Fig. 4.9). It differs from the "open-air" firing structures at Mehrgarh and Nausharo in that it is shallowly dug into a flat surface rather than placed on a sloping surface, it is very much smaller (80 x 75 cm, 30 cm deep), and it "has a definite opening to the west for air and possibly fuel" (Dales & Kenoyer 1991: 235). The bottom of the "tear-drop shaped" (ovate) pit contained a thick layer of ash, covered with a layer of potsherds. Dales and Kenoyer (1992:60) draw on modern potters' techniques to suggest that the sherds may have been used as "setters to hold and cover the objects being fired and also to protect them from direct contact with the fuel".

One of the most noteworthy aspects of this structure is the concentration of numerous low-fired triangular terracotta cakes and *mushtikas* ("potato shaped clay lumps with finger impressions") in the western opening ("mouth") of the pit (Dales & Kenoyer 1991:235, 1992:60). Dales and Kenoyer note that modern Pakistani potters place old vessels or stones at the mouth of similar firing structures to allow air into the structure while sealing in the heat. As there is no stone for hundreds of kilometers, Harappan potters (and cooks) seem to have substituted terracotta cakes and lumps of various shapes in their firing structures and hearths, although terracotta cakes are also found in hearths in stone-rich areas, such as Nausharo (e.g., C. Jarrige 1994). The cakes, lumps, and stones would also function to reduce fuel consumption and even out the supply of heat over the course of the firing, by absorbing heat early in the firing sequence, then radiating heat as the fuel was consumed. (Many thanks to my anonymous reviewer for pointing this out.) Ethnographically, the objects to be fired would be placed in the main part of the pit, behind the cakes and lumps, together with the fuel. And based on modern ethnographic analogies, Dales and Kenoyer (1991:235, 1992:60) suggest that the filled pit was then covered with mud to form a domed covering with vent holes, which was broken open and rebuilt after each firing. (No discovered fragments of such roofing are mentioned.) Finally, they note that "no complete objects were found in this kiln", but suggest that the size of the structure indicates the firing of only small vessels or figurines (Dales & Kenoyer 1992:60).

Single-Chamber Structures ("Ovens") from Mehrgarh

The vast majority of single-chamber firing structures are reported from the site of Mehrgarh, particularly from the southern half of the mound of Area MR.1. This region of Mehrgarh is incredibly rich in pottery manufacturing evidence, and includes numerous areas with unbaked pottery, small terracotta objects, potters' tools, storage rooms containing hundreds of vessels representing a wide variety of types, and numerous firing structures. Trash from the potters' areas contained huge quantities of overfired ceramics mixed with other debris, including pebbles used in the firing process. (C. Jarrige *et al.* 1995; Santoni 1989) I will focus here on firing structures from Sectors MR.F and K, in deposits dating to the early to mid-third millennium B.C. MR.F/K was both a craft and habitation area during Mehrgarh Period VII (very roughly *ca.* 2700 to 2500 B.C.). A trench through the mound also showed the presence of "intensive potters' activity" during the preceding Period VI (roughly *ca.* ?3000 to 2700 B.C.), including firing structures. (Santoni 1989:176; C. Jarrige *et al.* 1995:217,Fig. 4.21,427, see also 136,183 for firing structures from other parts of MR.1) The firing structures in MR.F/K ("ovens") are carefully distinguished from the roughly contemporaneous "large open firing areas" discussed above. Santoni (1989:176) states that these "ovens were probably used to fire the more elaborated kinds of pottery and particularly the painted grayware as well as small terracotta objects. Large jars and common pottery were fired in large open firing areas".

The Period VI deposits (roughly *ca.* ?3000 to 2700 B.C.) contain one of the most convincing examples of a definite pottery firing structure. It was rebuilt 3 times in the same location (unfortunately, no plan of the Period VI finds is available). All three re-buildings are described as a "free-standing kiln" or "cupola kiln", built of clay on top of a layer of small stones or stones and sherds (C. Jarrige *et al.* 1995:427; Santoni 1989:176;Fig. 2). The second rebuilding was the best preserved, and was a round domed structure (1.6 m external diameter, 1.2 m internal diameter, preserved to a height of 35 cm) built on a layer of stones and sherds placed inside the base of the first structure. It had mud-brick walls and (like all of the re-buildings) was lined with clay that had been heavily burnt during firing, grading from "grey-white nearest the heat source to dark red to a lighter red" (C. Jarrige *et al.* 1995:427). This second rebuilding was filled with clay, earth, and a few stones and sherds, but no fragments of a possible roof or dome. Small stones were placed on top of the filling to form the base of the third re-building. The structure was surrounded by ashes, broken wall fragments from former firing structures, and wasters, including large fragments of Faiz Mohammed painted grayware (Santoni 1989:176; C. Jarrige *et al.* 1995:427). Fragments of unfired clay vessels were also found in the deposits around the structures (C. Jarrige *et al.* 1995:428). Note that in all cases the stones were found <u>under</u> the firing structure base, forming part of the structure, not <u>inside</u> the structure, unlike many of the other firing structures at Mehrgarh where the stones seemed to have been part of the "kiln furniture" (see above and below).

The Period VII (roughly *ca.* 2700 to 2500 B.C.) firing structures in MR.F/K are more difficult to interpret. They are found in "yards" between or in front of rooms (C. Jarrige *et al.* 1995:217). It is not clear from the published descriptions whether all of these single-chamber "oven" or "firepit" structures consist of excavated depressions/pits or are built above a flat surface; my impression is that all of the Period VII structures are pits. They are mostly circular to ovate and usually only their bases were preserved, with maximum diameters ranging from

1.0 to 2.50 meters, with one extremely large exception (below). Almost all are described as having "well-smoothed sidewalls" (C. Jarrige *et al.* 1995:217), but none of them appear to have a "mouth" of any kind, unlike the "pit kiln" at Harappa (above). The total number of firing structures in MR.F/K is not specified, but there are at least six separate groups of structures mentioned for the Period VII deposits. At least two of these structures in MR.F/K were destroyed and rebuilt three times, in the same location but at different orientations and sometimes different sizes. (C. Jarrige *et al.* 1995:217,Fig. 4.21; Santoni 1989: 176,Fig. 1). This phenomenon of re-building is also seen for the "cooking hearths" excavated recently at Harappa. (Meadow & Kenoyer in press) In contrast, the double-chamber pottery kilns at Lal Shah (below) were seldom re-oriented or re-sized during restorations of a particular structure, emphasizing the greater effort needed to construct a double-chamber updraft structure.

It is tempting to re-interpret some of these features as hearths or ovens for cooking rather than pottery firing. For example, the "firepit" at Locus CI/Square 10A sounds remarkably like the "fire alters"/"firepits"/"cooking hearths" described in the Introduction, except that it does not have a small central block/pillar: an oval structure about 15 cm deep with fire-baked walls, filled with ashes and charcoal mixed with burnt pebbles, with broken terracotta cake fragments nearby. Another structure at Locus CXIII/Square 1B (1.15 by 0.9 m) contained charged seeds of wheat, barley, and grapes, in addition to ash and charcoal, which would tend to lend weight to a cooking function. However, modern potters in the region frequently use as fuel both animal dung (containing seeds) and agricultural wastes, such as incompletely winnowed straw from cereals. (C. Jarrige *et al.* 1995:218; see also Rye 1981:104) And at least one of the other structures in this area was heated to the point of vitrification, far above cooking temperatures. The lowest base of a 1.8 by 1.5 m structure re-built 3 times (in square 1C) had a bottom almost vitrified with the heat, and was filled by a large number of heat-cracked pebbles and several overburnt terracotta cakes/lumps, "some almost vitrified" (C. Jarrige *et al.* 1995:217). (Note that the last re-building of this structure (1.7 by 1.4 m) is the only one of the Period VII structures said to have a wall made of bricks.) Once again, it is clear that it is very difficult to determine the function of "single-chamber" or "firepit" structures of the Indus region, especially on the basis of shape characteristics alone, and "[i]t is quite probable that these kilns and firepits were used for various different purposes" (C. Jarrige *et al.* 1995:217).

Two of the remaining structures in MR.F/K Period VII are particularly interesting because they appear to incorporate transverse walls. Could these be a type of two-chamber kiln, with either the transverse wall supporting a suspended floor, or with the combustion and firing chambers separated horizontally rather than vertically? One of these structures (in Square 10B) is very oddly shaped, consisting of an almost L-shape divided into an oval and a semi-circle by a transverse, east-west wall. Only the semi-circular portion is floored with burnt pebbles, but no other information is given. (C. Jarrige *et al.* 1995:217,Fig. 4.21) There is more data for the other structure, which is the largest structure from MR.F/K, a 2.9 by 2.25 m oval with a hollow bottom and concave walls which "suggest the presence of a cupola" (C. Jarrige *et al.* 1995:217). This structure contained charcoal, charred seeds, and ashes, together with about thirty clay "sling balls", and "blocks" that may be fragments from the collapsed roof of the structure. Many of these fragments have mat or basket impressions (and in one case cloth), and Santoni suspected that mats or baskets were either used as a frame for a temporary roof, or for filling cracks or holes during firing. (C. Jarrige *et al.* 1995:217) No pottery wasters are reported from this structure, although a high level of heat is suggested due to the solidification of the ashes into whitish blocks (C. Jarrige *et al.* 1995: 217,Fig. 4.19A). About five courses of bricks are still preserved of a partition wall across the short axis of this structure, dividing off a third of the eastern end. It appears from the plan (Santoni 1989:Fig. 1) that the southeastern corner of the structure is missing, and that the wall does not completely stretch across the structure in the south, leaving both the western two-thirds and particularly the eastern third of the structure open from this point. However, it is not at all clear if this was the case in antiquity, or is due to post-use destruction. This transverse wall is only one brick thick, and does not seem substantial enough to support a suspended floor with a load of vessels. While the question of horizontally separated chambers remains, this wall may simply have supported the roof.

DOUBLE-CHAMBER UPDRAFT FIRING STRUCTURES (KILNS)

The last type of firing structure, the double-chamber updraft kiln, is the most common type of firing structure found (or at least recognized) at sites in the greater Indus region. (Figure 1) It is usually assumed to have been used for pottery firing. Most of the published examples come from Mohenjo-daro, but there are also well-described examples at Harappa, Lal Shah and Nausharo in the Kachi Plain. There are many more examples of firing structures in the literature, from excavations at Amri (Casal 1964), Balakot (Dales 1974), Kot Diji (Khan 1965), Lothal (Rao 1979), and several sites in present-day India (Sharma 1979-1989); and from surfaces surveys at Chanhu-daro (Sher & Vidale 1985; Vidale 1989) and sites in Cholistan (Mughal 1984, 1990). Many of these have a central pillar, and are thus presumably double-chambered. The structure from Balakot was particularly well preserved, and may have been used for firing terracotta figurines (Dales 1974).

One of the great difficulties in analyzing firing structures from the Indus is discriminating between double-chamber and single-chamber structures on the basis of the archaeological remains. Most examples only have the combustion chamber preserved, and rarely a suspended floor, but are assumed to be double-chamber updraft firing structures because of the presence of a substantial central pillar. This assumes a certain type of double-chamber kiln structure—see the "Cylindrical Kiln" section below for suggestions about a possible double-chamber updraft kiln <u>without</u> a pillar. But the variations in double-chamber updraft kilns could form a paper in themselves, even with the scanty data available for most of the Indus firing structures. In this section I will only describe the remains of the double-chamber updraft firing structures found in recent excavations at Lal Shah, some of the most thoroughly studied of the recently excavated double-chamber structures, then briefly mention some of the similar firing structures at Harappa, Mohenjo-daro, and other sites.

Double-Chamber Updraft Kilns at Lal Shah, Mehrgarh

The remains of six double-chamber updraft firing structures, dating to two different phases of use, were found at Lal Shah, an 'industrial site' near Mehrgarh Area 1 of Mehrgarh Period VIIC (2700-2500 B.C.) (Pracchia 1985:460; Quivron 1994:629; C. Jarrige *et al.* 1995:85,462,506). These specialized production sites, with no evidence for associated habitation, are a phenomenon that is found most often during the "Mature Harappan" phase (*ca.* 2600 to 2000/1900 B.C.), according to survey work done in the desert regions between India and Pakistan where many such specialized firing sites were identified (Mughal 1984, 1990). The Lal Shah firing structures are all elongated "tear-drop" shaped structures with a central pillar. Based on measurements taken from Pracchia's drawing (1985: Fig 2), the Lal Shah structures have length to width ratios of roughly two to one (including the fire tunnel). Very roughly, the two earlier phase kilns, Kilns 1 and 2, are the largest at not quite 3 meters by 1-1/2 meters. Kilns 3 and 6 are about 2 meters by 1 meter and Kiln 4 just less than this in both directions, while Kiln 5, the roundest, measures roughly 2 meters by 1-1/4 meters.

Kiln 1 (Wright's Kiln A) was initially excavated by Wright (ms), then Pracchia (1985) conducted further excavations over 120 square meters in 1985. Kiln 1 was the only kiln still to have the base of the firing chamber (grate/suspended floor) preserved. It consisted of a combustion chamber about 1.5 m in diameter, cut into the side of the hill and lined with a single course of bricks laid edgewise (10 cm thick). A 50 cm diameter pillar built of baked bricks and clay filler supported a grate/suspended floor made of bricks stretched between the central pillar and the walls. The combustion chamber was finally completely plastered with vegetal (straw)-tempered clay, including the pillar and the domed underside of the grate/ suspended floor. There were three layers of this vegetal-tempered clay lining, indicating that the combustion chamber had been used and re-plastered on at least three occasions. At least the final use of the kiln reached temperatures high enough to result in the slagging of this lining (below). (Pracchia 1985:460, Fig. 3; Wright ms; Wright & Mishara ms; C. Jarrige *et al.* 1995:462,Fig. 10.30)

The remains of the grate/suspended floor contained 7 cone-shaped holes (narrower in diameter on the firing chamber end than the combustion chamber end) placed around the perimeter of the grate (Wright ms). The original structure probably had about 12 of these interior flues, allowing hot air circulation between the lower combustion chamber and the firing chamber above it. As is usual for updraft kilns (Rye 1981:100), the area of maximum heat during operation was in the combustion chamber around these interior flues, as shown by the formation of stalactites of molten clay (C. Jarrige *et al.* 1995:462). The wall of the firing chamber, which originally stood above ground level, was made of bricks and preserved to a height of 25-30 cm. The floor of this chamber (the upper side of the grate) was coated with 3 layers of a white coating (discussed below), again representing at least three occasions of use. The upper part of the firing chamber may have been impermanent, such as a covering of potsherds, earth, mud plaster, and/or the circular flat straw-tempered clay disks mentioned above, which would be replaced for each firing. This would make loading the kiln much easier. (Wright ms; Pracchia 1985:460; C. Jarrige *et al.* 1995:462,Fig. 10.30; Santoni 1989:176)

Unfortunately, the fire tunnel (Wright's "fire box") was cut by a later grave pit (Pracchia 1985:Fig. 3). Enough remained to show that while some ash and charcoal were found within the combustion chamber, most of it was found piled near or within the fire tunnel, in several layers to a depth of 25-30 cm (Wright ms). One of the most unusual features of the structure is a "large hole" near the junction of the fire tunnel with the combustion chamber. This "hole" connects the combustion chamber with the exterior of the kiln. Wright suggests that the circulation of heat within the kiln was controlled by manipulating both this "regulating hole" (which would also give the potter a view of conditions in the combustion chamber) and the stoking hole (i.e., the entrance to the fire tunnel). (Wright ms; Santoni 1989:176; C. Jarrige *et al.* 1995:462)

Subsequent analyses of the slagged clay lining from Kiln 1's combustion chamber using SEM and X-ray diffraction indicate that it was heated to above 800°C (Wright & Mishara ms). The white coating from the grate/suspended floor of the firing chamber was also analyzed by SEM and XRD, and found to be impure gypsum, not lime as was previously thought (Wright & Mishara ms; Jarrige *et al.* 1995:462; Pracchia 1985:460). Although traces of possible post-depositional gypsum salts were found throughout the structure, these abundant coating layers were probably deliberately applied rather than post-depositional (Wright & Mishara ms). Wright and Mishara (ms) suggests that the gypsum might be from secondary use of the kiln for gypsum plaster manufacture (normally fired at 100-200°C), as well as pottery firing at higher temperatures. It is also possible that the gypsum coatings were deliberately applied, and had some sort of refractory or pyrotechnic function, to protect the firing chamber or the vessels being fired from collapse or sticking during high temperatures. (Cp. the white coatings—in this case steatite/talc paste—on "dishes" from Harappa (Miller in press).) Either scenario would explain the regular re-coating of the grate.

While the remaining kilns were preserved to lesser degrees (down to Kiln 6, with only traces of the base of the combustion chamber left), all had both the vegetal-tempered clay linings of the combustion chamber walls and the white (gypsum) coating of the grate regularly replaced (Pracchia 1985:460). The regular restoration of "calcareous lime-like pavings" around the kilns allowed the chronological division of the use of the six kilns into two structural phases (Pracchia 1985:460,466-467). (Based on the description, the "calcareous lime-like pavings" are probably also gypsum.) Kilns 1 and 2, the most complete kilns, were operating simultaneously during the first phase, during which 7-8 restorations of the kilns were noted. The remaining four kilns were operating during the second phase, and there were signs of at least two restorations of each of these four kilns during this second phase. Pracchia (1985:466,ftnt. 1) notes that if these restorations corresponded to a yearly cycle of firing activity, the area could have been used for a period of 9 to 14 years, with or without a break between the two phases.

Firing waste products ("shards of cracked or over-fired pottery, setters and supports for earthenware jar manufacture") covered the entire excavated area in varying concentrations (Pracchia 1985:460). (Pracchia (1985:460-461) and C. Jarrige *et al.* (1995:85,462,506) discuss the pottery types found in association with the kilns.) The only kiln furniture found were fragments of setters, but these were a bit different from the "rectangular-sectioned bangles" thought to have been used as setters from Mehrgarh Area 1, MR.K (Santoni 1989:181,184-185). Based on the photograph (Pracchia 1985:Fig. 7), the Lal Shah setters are short, thick, roughly-made, semi-circular coils of clay mixed with straw, most about 3-5 cm in cross-section. Finds of "conical stands used for jar forming" (molds or chucks), two smoothers made from pottery sherds, and a bull figurine in unbaked clay (Pracchia 1985:461,Fig. 6,8,9), indicate that this area may have been used for forming as well as firing of pottery and other terracotta objects.

It is also of note that three of the Lal Shah kilns (2, 5 and 6, the westernmost kilns) are oriented roughly north-south (with the fire tunnel opening to the south), while the remaining three kilns are all oriented in very different directions, both from north-south and from each other. (Pracchia 1985:Fig. 2; the implication in C. Jarrige *et al.* 1995:462 that Kiln 1's mouth opens southwards is incorrect—it opens northwest. Kiln 2's mouth opens southwards.) A south-facing orientation may have related to prevailing winds during the summer season (when modern firings normally take place). In any case, there was clearly no concern for at least Kilns 1, 3, and 4 with either the direction of prevailing winds or with conformation with culturally important directions. (Habitation structures and streets are usually oriented roughly north-south at the planned sites of the Indus civilization.) Since 1 and 2 are the earliest kilns, and 5 and 6 probably the latest, it may be that making structural use of the small mound into which Kilns 1, 2, 3 and 4 were dug was more important than other considerations of orientation. Orientation north-south then again became a primary concern for Kilns 5 and 6, when the space around the small mound was filled.

Double-Chamber Updraft Kiln at Harappa

The remains of a large double-chamber updraft structure, Kiln 100, was found on the northwest corner of Mound E at Harappa, one of three firing structures found in an area of some 70 square meters (Dales & Kenoyer 1991, 1992; Wright 1991). (See the "pit kiln" described above.) Its last use is radiocarbon dated to about 2300 B.C. (calibrated) (Dales & Kenoyer 1991:43,Fig. 4.9). Like the kilns at Lal Shah, Kiln 100 is "tear-drop" shaped, but it is larger and less elongated than any of the kilns at Lal Shah, measuring 3 m in length (including the fire tunnel) by 2 m in width. Only the bottom meter of the combustion chamber was preserved, including the large central pillar (about 1 m diameter) (Kenoyer 1994:Fig 29.3). All of the interior walls of the combustion chamber were heavily vitrified, including the pillar. Sherds (mostly unvitrified) from the surrounding deposits were analyzed by Wright, and seem to indicate that only the most complex types of vessels in the assemblage of this period (Harappa Period 3) were being fired in this firing structure, including very large storage jars of a type traded as far as Oman. Wright also notes that "while some misfired pottery and firing structure wasters have been found, they are not present in large quantities, suggesting a well-developed knowledge" of firing (Wright 1991:83). Pottery production tools found near the kiln include a chuck/mold, hematite, bone and stone tools, and clean clay (Dales & Kenoyer 1991:235, 1992:62; Wright 1991:78-83).

The most unusual feature of Kiln 100 is a type of clay slab structural element, always heavily vitrified on its concave side, most of which were found in the fire tunnel or the entrance to the combustion chamber. Several functions have been suggested for these elements—as flues (either external or between the combustion chamber and the firing chamber), as bars to form the grate/ suspended floor, or as arch elements for the roof of the fire tunnel. Kenoyer (1994:352) has used the last suggestion in his reconstruction of Kiln 100 for experimental firings.

Double-Chamber Updraft Kilns at Mohenjo-daro

Other examples of similar firing structures come from the excavations at Mohenjo-daro in the 1920s & 1930s, particularly from Mackay's excavations in DK-G area. (Mackay 1938, Marshall 1931) Some seven possible double-chamber updraft kilns were found in deposits of the "Late II and I (b & a) Phases", one in the northern portion of DK-G, one in the middle of the street dividing DK-G, and five in the southern portion. Dating of these structures is a serious problem, as the great majority of the excavations at Mohenjo-daro were done prior to radiocarbon dating. (See Franke-Vogt 1994 for discussion.) The most helpful aid to dating these old excavations will be further development of a pottery chronology for Mohenjo-daro, but it is worth noting that two of the DK-G firing structures had associated wasters of numerous 'Pointed Base Goblets' (among other vessels). (Mackay 1938:54,103) This is a very distinctive pottery type which at Harappa is primarily restricted to Harappa Period 3C (*ca.* 2250 to 2000/1900 B.C.) (Jenkins 1994:325-326; Meadow & Kenoyer in press).

Little information is given for most of the DK-G firing structures other than their measurements; the presence of a pillar is not always mentioned even in kilns with portions of a grate/suspended floor or pillar clearly visible in the photographs. Four of the seven definitely had pillars and/or suspended floors preserved. But since some of the structures have neither published photographs nor drawings, I can only guess that all but one were probably double-chamber updraft kilns, based on their grouping together by Mackay. The exception is an enormous, "not perfectly circular" structure some 4.8 m by 4.7 m, with a 46 cm thick, "very roughly built" wall of broken and unbroken bricks of various sizes (preserved to a height of 1.5 m), which was found in an open space between Blocks 7 & 9. Described as "unfinished", it may be the remains of a single chamber/open air type of firing structure, although Mackay (1938:84-85) notes that two short parallel walls to the south of the structure may be the remains of a fire tunnel, which is characteristic of either double-chamber updraft structures, or the "pear-shaped/sub-triangular" single-chamber structures.

The remaining six structures from DK-G are described as irregular ovals or ellipticals or "practically circular", with length to width ratios, where given, of 2.0 m to 1.0 m, 1.8 m to 1.4 m, 2.2 m (diameter), and 1.4 m by ? m. (It is not clear whether the fire tunnel is included in these measurements.) They seem to have mostly been built of broken (baked?) bricks and plastered with clay/mud on their interiors, including the pillars built also of (baked?) bricks. (Mackay 1938:33,62, Pl. Lb, 102-103,Pl. XXIII2,Pl. Ld,154,Pl. Xd,Pl. XX) At least two of these structures had portions of their clay grates/suspended floors preserved, perforated by holes (interior flues) averaging 10 cm and 12 cm in diameter arranged around the edge of the grate (Mackay 1938:62,102). The best preserved kiln had almost a complete grate/suspended floor some 12-13 cm thick, with the holes in a ring around the edge of the grate plus a singe hole in the center, offset from the supporting pillar below (Mackay 1938:102,Pl. XXIII2,Pl. Ld).

Three other double-chamber updraft kilns found in earlier excavations in other areas of Mohenjo-daro (Marshall 1931:193,Pl. LIa,226,Pl. LVIIIb) and by the IsMEO surveys (Pracchia 1987:153; Pracchia *et al.* 1985:219) are slightly different in shape, with a much more abrupt join between the body of the firing structure and the fire-tunnel. At least two of the three had "tongue-support" type of pillars, projecting out from the rear wall of the kiln (Marshall 1931:Pl. LIa; Pracchia 1987:153), rather than the more common free-standing central pillars. Whether this reflects differences in use of the kilns or chronological changes in kiln construction cannot be determined on the basis of these examples. But obviously, there were probably a variety of types of double-chamber updraft firing structures in use, and we are only beginning to distinguish between them. Two more types of structures which may have been variants of double-chamber updraft kilns for pottery firing will be briefly discussed below.

"Pear-Shaped" (Brick-Lined) Firing Structures

Vats' (1940) excavations at Harappa uncovered 16 firing structures of various types in an area of less than 60 x 60 m (Trench IV on the southwest corner of Mound F). These 16 structures were found in four different major phases, and it is not at all clear how many were actually in operation at once. (At the moment, these four phases can only be vaguely dated as "Mature Harappan", *ca.* 2600 to 2000/1900 B.C.) However, the majority of these structures were of a very unusual type. 13 pear-shaped firing structures were found, eight of which were lined with baked brick, and 5 of which were "mere pits dug in the ground". Vats (1940:472) only gives general size measurements, indicating that the structures ranged from 1.9 m to 2.5 m in length, and from 1.0 m to 1.7 m in width, and were preserved to no greater height than about 75 cm. (Figure 2)

All of the structures resemble each other in shape (Vats 1940:472-473), and are similar in outline to the "tear-drop" shaped double-chamber updraft kilns found at so many Indus sites. However, only three of Vats' structures (Fk, Fn, and Fc) had central pillars, one of which projects out from the rear wall; the remaining 10 firing structures did not have pillars at all. Vats did not believe that the two free-standing pillars were used to support the floor of another chamber. However, Vats (1940:473-474) mentions a brick-lined firing structure found by Mackay at Mohenjo-daro that is an "almost exact parallel of the pear-shaped furnaces from Harappa", which clearly has a free-standing central pillar (Mackay 1938:103,Pl. Ld). Mackay notes that this structure was exposed to high temperatures, so that the pillar is partly vitrified. Another very similar "pear-shaped", brick-lined, mud-plastered structure from Mohenjo-daro (Mackay 1938:62,Pl. Lb) even has the suspended floor preserved, with holes around its perimeter.

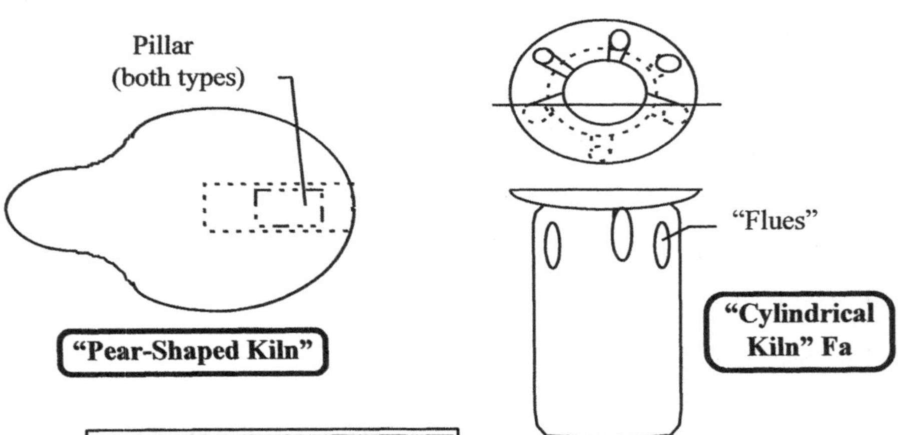

FIGURE 2 (after Vats 1940)

Perhaps the most intriguing aspect of these firing structures from Harappa is that all of the brick-lined firing structures, and probably at least some of those not brick-lined, are plastered with "mud mixed with a quantity of sand" ("to resist fusibility") (Vats 1940:472). "In several instances the original mud-and-sand plaster, which was generally turned into slag" was re-plastered, indicating that at least some of the firing structures were maintained and re-used. All of the other firing structures which I've seen or for which I have information, other than cylindrical "Kiln Fa" below, are plastered with straw-tempered clay—sand tempering is apparently unique to the structures on Mound F at Harappa. (Mackay makes no mention of any temper in the plaster on the similar structures from Mohenjo-daro.)

Vats briefly lists the possible products fired in these firing structures, based solely on the size of the structures and the high temperatures reached; there are no 'wasters' of any kind of product mentioned from this area of Mound F. He (1940:473) particularly dismisses pottery firing, as the structures would be "inconveniently small" and as "pottery does not require such intense heat as is evidenced by the vitrification in some of these furnaces". But the size of these firing structures is quite similar to known pottery firing structures from other excavations (above), and many of these structures are heavily vitrified. In fact, Mackay (1938:103) is quite insistent that the small kiln from Mohenjo-daro was for pottery firing. Without any discussion of mis-fired products, it is extremely difficult to assign a function to these structures, but the firing of pottery and/or other terracotta objects certainly cannot be ruled out. (NOTE: As is discussed in detail in Kenoyer and Miller (in press) the widespread notion found in the secondary literature that the 16 firing structures found on Mound F were associated with copper/bronze working is incorrect.)

"Cylindrical" Firing Structure

In his excavations on Mound F, Vats (1940:470-472,53,54) also found a "cylindrical kiln", Fa (in Stratum II, see above re dating). (Figure 2) Portions of Fa are still in place on Mound F at Harappa and I examined this structure in detail during recent conservation work (Miller in Kenoyer & Meadow 1992). When first excavated, half of the original structure was preserved (split vertically), revealing a surface so heavily vitrified that "its mud walls have actually run down in pencil-like formations of slag" (Vats 1940:471). Vats describes the structure as a pit 1.0 m in diameter and 1.1 m deep (although excavated to a depth of 1.6 m, there was no trace of fire below the 1.1 m depth). It was at least partially constructed with mud-brick, and its interior was covered with a sand-tempered mud plaster (unpublished field notes, Miller in Kenoyer & Meadow 1992). Based on structural considerations and my examinations of the remaining walls of Fa, the existing structure was probably at least semi-subterranean. Vats (1940:470-471,Fig. d,Pl. XVIIa,b) notes that it had a vaulted roof, part of which was broken and lying inside the kiln, with four "flues" remaining at the springing of the vault and two more conjectured. Vats (1940:472) suggests that these "flues" were outlets for smoke and/or inlets for fuel that could be closed during the operation of the kiln. In addition, a fifth, slanting "flue" (barely visible in the photographs and not shown in the schematic) entered the kiln at "an obtuse angle" lower down, apparently just above the lowest vitrified levels of the kiln. Vats (1940:471-472) proposes that this lower "flue" was "used as an air channel worked by bellows from above."

However, in the photographs and in the schematic drawing (Vats 1940:470,Fig. d, Pl. XVIIa,b), there is a shallow depression around these flues, with a diameter of perhaps 1-1/4 to 1-1/2 meters, which is not mentioned in the text. It is possible that Fa was a double-chamber updraft kiln, and these "flues" were air ducts leading into an upper firing chamber which was larger than the combustion chamber below. The "vaulted roof" would be the remains of the grate/suspended floor of the kiln, which was <u>not</u> supported by a central pillar due to the smaller size of the combustion chamber. The lower "flue" might be the remnants of a "regulation hole" similar to one described for Kiln 1 at Lal Shah (above). If the preserved part of Fa were indeed only the combustion chamber, it would explain the extreme heat indicated by the vitrification of the walls, as the combustion chamber and fire tunnel are usually the most vitrified portion of a double-chamber updraft kiln (Rye 1981:100). However, it would be necessary to do experimental recreations to see if this proposed reconstruction is structurally possible.

CONCLUSION

I have emphasized that there is a continuum in the "types" of firing structures in use in the Indus. Shallow "pit" kilns with built-up sides grade into single-chamber structures with impermanent upper portions and roofs. Repeatedly used "open-air" structures with plastered clay coverings grade into rapidly built single-chamber structures using loose mud bricks or clay lumps, both employing slopes and simple construction techniques to encourage a draft. The Mohenjo-daro single-chamber structure shows how these same principles can be employed in increasingly more permanent structures. We do not seem to see a gradation in double-chamber structures in the Indus (yet); the only structures with the separation of fuel and objects fired are well-developed double-chamber updraft kilns, unless the single-chamber "ovens" with a dividing wall indeed functioned as "horizontal" double-chamber kilns. With the current data set, double-chamber updraft kilns appear to be a phenomenon of the urban phase of the Indus civilization, first used around 2600/2500 B.C. at the beginning of this urban or "Mature Harappan" period. However, both the analysis of firing structures and the general Indus chronologies need to be considerably refined before we can discuss the origins and development of double-chamber updraft kilns for the Indus region.

The other point stressed is that there is not an evolutionary progression of firing structure types. We do see the development of new, increasingly complex firing structures over time, but variants of the older structures apparently continue to be used along with the new "types". (For example, the contemporaneous use of single-chamber and open-air structures at Mehrgarh, and of double-chamber updraft and open-air structures at Nausharo.) Why this is occurring is one of the more interesting questions about Indus pottery firing technology. How did the Indus craftspeople choose which type of firing structure to use? Were their decisions dependent on the quality of the objects being produced? Or the size of the objects? Or the fuel costs? To contrast the costs of a double-chamber updraft kiln with an "open-air" firing structure, the updraft kiln requires more labor to build and maintain, requires the permanent dedication of space to firing (an important issue in urban contexts), and uses more expensive fuel, as most updraft kilns do not use agricultural waste or dung, but primarily wood. The advantages of an updraft kiln are that the firing can be more easily controlled, is less affected by weather conditions, higher temperatures can be reached, and the products are protected from the smoke and dirt of the fuel. Which of these issues were important for Indus potters? High firing temperatures do not seem to have been a problem, but the separation of products from fuel and the maintenance of an oxidizing atmosphere may have been important. For example, Jenkins (1994:324-325) notes that the "Early Harappan" (Harappa Period 2, roughly *ca.* 3000/2800 to 2600 B.C.) pottery assemblages at Harappa are often unevenly fired and fire-clouded, while the "Mature Harappan" (Harappa Period 3, *ca.* 2600 to 2000/1900 B.C.) vessels are more evenly fired. Also, the earlier sherds have a "mottled, gray to brown color, indicating that they were probably fired in a slightly reducing atmosphere," while the Period 3 pottery is generally a light red and fired in an oxidizing atmosphere. The ability to fire in a wider range of weather conditions may also have been important. With the increase of vessel mass-production during the urban or "Mature Harappan" phase, it may have been necessary to fire pottery more often, almost year-round rather than only during the dry season. Again, to answer these questions we need better chronological control; we are still working at the +/- 100 year level in the Indus, so that it is difficult to truly discuss "contemporaneity" and "development".

Finally, one of the reasons why we find so many double-chamber updraft kilns is because they are hard to miss, having been subjected to such high temperatures that the combustion chambers have vitrified. Why are there so many vitrified structures, especially the double-chamber updraft kilns? Occasional environmental problems, like windstorms, seems unlikely for such a large number of structures. Indus pottery is not particularly high fired (roughly 800°C, on average). And while there are many overfired pottery wasters at Mohenjo-daro (Pracchia *et al.* 1985), there are surprising low amounts of overfired and vitrified pottery wasters at Harappa (Miller, dissertation in progress; Wright 1991:83). How do we resolve this enigma of vitrified kilns and few wasters? Part of the answer may be the high illite content of Indus region clays, causing them to vitrify at relatively low temperatures (around 1000°C). To fire a large load of objects to 800°C in the upper chamber of a large kiln (itself a major heat sink), it would probably have been necessary to generate temperatures of over 1000°C in the combustion chamber. The Indus potters, like Indus craftspeople working with other materials, must have continuously pushed the limits of what they could achieve.

ACKNOWLEDGMENTS

Many thanks to Rita Wright for access to her unpublished research reports and notes, and to the Harappa Archaeological Research Project, especially J.M. Kenoyer, for access to their unpublished field notes. I especially appreciated and benefited from the thoughtful and extensive remarks of an anonymous reviewer, and the editorial aid of Prudence Rice.

REFERENCES

Anderson-Gerfaud, P., M.-L. Inizan, M. Lechevallier, J. Pelegrin, and M. Pernot
1989 Des lames de silex dans un atelier de potier harappéen. Interaction de domaines techniques. *Comptes Rendus de l'Académie des Sciences,* series 2, 308:443-449.

Audouze, F., and Catherine Jarrige
1979 A Third Millennium Pottery-Firing Structure at Mahrgarh and Its Economic Implications. In *South Asian Archaeology 1977*, edited by Maurizio Taddei, pp. 213-221. Istituto Universitario Orientale, Naples.

Casal, Jean-Marie

1964 *Fouilles d'Amri.* Publications de la Commission des Fouilles Archéologiques, Fouilles du Pakistan. Librairie C. Klincksieck, Paris.

Dales, George F.

1974 Excavations at Balakot, Pakistan, 1973. *Journal of Field Archaeology* 1:3-22.

Dales, George F. and Jonathan Mark Kenoyer

1986 *Excavations at Mohenjo-daro, Pakistan: the Pottery.* University Museum Monograph 53. University Museum, Philadelphia.

1991 Summaries of Five Seasons of Research at Harappa (District Sahiwal, Punjab, Pakistan), 1986-1990. In *Harappa Excavations 1986-1990: A Multidisciplinary Approach to Third Millennium Urbanism,* edited by Richard H. Meadow, pp. 185-262. Prehistory Press, Madison, WI.

1992 Harappa 1989: Summary of the Fourth Season. In *South Asian Archaeology 1989,* edited by Catherine Jarrige, pp. 57-67. Prehistory Press, Madison, WI.

Franke-Vogt, Ute

1994 The 'Early Period' at Mohenjo-daro. In *From Sumer to Meluhha: Contributions on Southwest and South Asian Archaeology in Memory of George F. Dales,* edited by J. M. Kenoyer, pp. 27-49. Wisconsin Archaeological Reports No. 3, Department of Anthropology, UW-Madison, Madison, WI.

Jarrige, Catherine

1994 The Mature Indus phase at Nausharo as seen from a block of Period III. In *South Asian Archaeology 1993, Volume I,* edited by Asko Parpola & Petteri Koskikallio, pp. 281-294. Annales Academiae Scientiarum Fennicae B 271. Suomalainen Tiedeakatemia, Helsinki.

Jarrige, Catherine, and F. Audouze

1979 Etude d'une aire de cuisson de jarres du IIIe millénaire: Comparaisons avec les techniques contemporaines de la plaine de Kachi au Baluchistan. In *L'archéologie de l'Iraq, du début de l'époque néolithique à 33 avant notre ère,* edited by M.T. Barrelet, pp. 85-98. Paris.

Jarrige, Catherine, J.-F. Jarrige, Richard Meadow, & Gonzague Quivron, editors

1995 *Mehrgarh: Field Reports 1974-1985 from Neolithic Times to the Indus Civilization.* Dept. of Culture & Tourism, Government of Sindh, Pakistan, in collaboration with the French Ministry of Foreign Affairs, Karachi, Pakistan.

Jarrige, J.-F.

1979 Excavations at Mehrgarh-Pakistan. In *South Asian Archaeology 1975,* edited by J.E. van Lohuizen-de Leeuw, pp. 76-87 & Pl. 30-44. E.J.Brill, Leiden.

1995 Introduction. In *Mehrgarh: Field reports 1974-1985 from Neolithic Times to the Indus Civilization,* edited by C. Jarrige, J.-F. Jarrige, Richard Meadow, & Gonzague Quivron, pp. 51-103. Dept. of Culture & Tourism, Government of Sindh, Pakistan, in collaboration with the French Ministry of Foreign Affairs, Karachi, Pakistan.

Jenkins, Paul Christy

1994 Continuity and change in the ceramic sequence at Harappa. In *South Asian Archaeology 1993, Volume I,* edited by Asko Parpola & Petteri Koskikallio, pp. 315-328. Annales Academiae Scientiarum Fennicae B 271. Suomalainen Tiedeakatemia, Helsinki.

Kenoyer, Jonathan Mark

1994 Experimental studies of Indus Valley technology at Harappa. In *South Asian Archaeology 1993, Volume I,* edited by Asko Parpola & Petteri Koskikallio, pp. 345-362. Annales Academiae Scientiarum Fennicae B 271. Suomalainen Tiedeakatemia, Helsinki.

Kenoyer, Jonathan Mark and Richard Meadow

1992 ms "Harappa Archaeological Project 1992: End of Season Report." UC-Berkeley, Harvard University, and UW-Madison. Unpublished manuscript submitted to the Pakistan Department of Archaeology.

Kenoyer, Jonathan Mark and Heather M.-L. Miller

In press Metal Technologies of the Indus Valley Tradition in Pakistan & Western India. In *The Archaeometallurgy of the Asian Old World,* edited by Vincent C. Pigott. Univ. of Pennsylvania Press, Philadelphia.

Khan, F.A.

1965 Excavations at Kot Diji. *Pakistan Archaeology* 2:11-85.

Lal, B.B.

1979. Kalibangan and the Indus Civilization. In *Essays in Indian Protohistory,* edited by D.P. Agrawal and Dilip K. Chakrabarti. pp. 65-97. B.R. Publishing Co., Delhi.

Mackay, Ernest J.H.

1938 *Further Excavations at Mohenjo-daro.* Govt. of India Press, New Delhi.

Marshall, John (editor)

1931 *Mohenjo-daro and the Indus Civilization.* Arthur Probsthain, London.

Meadow, Richard H. and J.Mark Kenoyer

In press. Excavations at Harappa 1994-1995: New perspectives on the Indus script, craft activities, and city organization. To be published in *South Asian Archaeology 1995,* edited by B. Allchin. Oxford & IBH, New Delhi.

Méry, Sophie

1994 Excavation of an Indus potter's workshop at Nausharo (Baluchistan), Period II. In *South Asian Archaeology 1993, Volume II,* edited by Asko

Parpola & Petteri Koskikallio, pp.471-482. Annales Academiae Scientiarum Fennicae B 271. Suomalainen Tiedeakatemia, Helsinki.

Miller, Heather M.-L.

In press. Locating Ancient Manufacturing Areas: High-Temperature Manufacturing Debris from Surface Surveys at Harappa, Pakistan. To be published in *South Asian Archaeology 1995,* edited by B. Allchin. Oxford & IBH, New Delhi.

Mughal, M. Rafique

1984 Recent Archaeological Research in the Cholistan Desert. In *Harappan Civilization: A Contemporary Perspective,* edited by Gregory L. Possehl, pp. 85-95. Oxford and IBH Publishing, Bombay.

1990 The Harappan Settlement Systems and Patterns in the Greater Indus Valley (Circa 3500-1500 B.C.). *Pakistan Archaeology* 25: 1-72.

Pracchia, Stefano

1985 Excavations of a Bronze-Age Ceramic Manufacturing Area at Lal Shah, Mehrgarh. *East and West* (new series): 35(4): 458-468.

1987 Surface Analysis of Pottery Manufacture Areas at Moenjodaro. The Season 1984. In *Interim Reports Volume 2,* edited by Michael Jansen and Günter Urban, pp. 151-166. German Research Project Mohenjo-Daro, RWTH Aachen and IsMEO, Roma; Aachen.

Pracchia, Stefano, Maurizio Tosi and Massimo Vidale

1985 On the Type, Distribution and Extent of Craft Industries at Moenjo-daro. In *South Asian Archaeology 1983,* edited by Janine Schotsmans and Maurizio Taddei, pp. 207-247. Istituto Universitario Orientale, Naples.

Quivron, Gonzague

1994 The pottery sequence from 2700 to 2400 BC at Nausharo, Baluchistan. In *South Asian Archaeology 1993, Volume II,* edited by Asko Parpola & Petteri Koskikallio, pp. 629-644. Annales Academiae Scientiarum Fennicae B 271. Suomalainen Tiedeakatemia, Helsinki.

Rao, S.R.

1979 *Lothal, A Harappan Port Town (1955-62), Vol. 1.* Archaeological Survey of India, New Delhi.

Rice, Prudence

1987. *Pottery Analysis: A Sourcebook.* University of Chicago Press, Chicago.

Russell, Nerissa

1995 The Bone Tool Industry at Mehrgarh and Sibri. In *Mehrgarh: Field reports 1974-1985 from Neolithic Times to the Indus Civilization.* Edited by C. Jarrige, J.-F. Jarrige, Richard Meadow, & Gonzague Quivron, pp. 583-613.

Dept. of Culture & Tourism, Government of Sindh, Pakistan, in collaboration with the French Ministry of Foreign Affairs, Karachi, Pakistan.

Rye, Owen S.

1981 *Pottery Technology: Principles and Reconstruction.* Manuals on Archeology No. 4. Taraxacum Press, Washington, D.C.

Santoni, Michelle

1989 Potters and Pottery at Mehrgarh During the Third Millennium BC (Per. VI & VII). In *South Asian Archaeology 1985,* edited by Karen Frifelt and Per Sørensen, pp. 176-185. Scandinavian Institute of Asian Studies Occasional Papers No. 4. Curzon Press, London & Riverside Press, Maryland, MD.

Sharma, D.V.

1979-1989 Harappan Furnaces. *Puratattva: Bulletin of the Indian Archaeological Society.* 11: 125-127.

Sher, G.M. and Massimo Vidale

1985 Surface Evidence of Craft Activity at Chanhu-daro, March 1984. *Estratto da Annali dell'Istituto Universitaro Orientale (Napoli).* 45: 585-598 + plates.

Sinopoli, Carla M.

1991. *Approaches to Archaeological Ceramics.* Plenum Press, New York.

Thapar, B.K.

1973. New traits of the Indus civilization at Kalibangan: an appraisal. In *South Asian Archaeology (1971)* edited by Norman Hammond, pp. 85-104. Noyes Press, Park Ridge, NJ.

Vats, Madho Sarup

1940. *Excavations at Harappa: Being an account of archaeological excavations at Harappa carried out between the years 1920-21 and 1933-34.* Government of India Press, Delhi.

Vidale, Massimo

1989 Specialized Producers and Urban Elites: on the Role of Craft Industries in Mature Harappan Urban Contexts. In *Old Problems and New Perspectives in the Archaeology of South Asia,* edited by Jonathan Mark Kenoyer, pp. 171-181. Wisconsin Archaeological Reports No. 2. Anthropology Dept., Univ. of Wisconsin-Madison, Madison, WI.

Wright, Rita P.

1985 Technology and Style in Ancient Ceramics. In *Ceramics and Civilization Volume I: Ancient Technology to Modern Science,* edited by W.D. Kingery, pp. 5-25.

1986 The Boundaries of Technology and Stylistic Change. In *Ceramics and Civilization Volume II: Technology and Style,* edited by W.D. Kingery, pp. 1-20. American Ceramic Society Inc., Columbus, Ohio.

1989a The Indus Valley and Mesopotamian Civilizations: A Comparative View of Ceramic Technology. In *Old Problems and New Perspectives in the Archaeology of South Asia,* edited by J.M. Kenoyer, pp. 145-156. Wisconsin Archaeological Reports Volume 2. Anthropology Dept., Univ. of Wisconsin-Madison, Madison, WI.

1989b New Perspectives on Third Millennium B.C. Painted Grey Wares. In *South Asian Archaeology 1987,* edited by Karen Frifelt, pp. 137-149. Scandinavian Institute of Asian Studies Occasional Papers Number 4. Curzon Press Ltd., London.

1989c New Tracks on an Ancient Frontier. In *Archaeological Thought in America,* edited by C.C. Lamberg-Karlovsky, pp. 268-279. Cambridge University Press, Cambridge.

1991 Patterns of Technology and the Organization of Production at Harappa. In *Harappa Excavations 1986-1990: A Multidisciplinary Approach to Third Millennium Urbanism,* edited by Richard H. Meadow, pp. 71-88. Prehistory Press, Madison, WI.

ms Lal Shah Kiln Excavations. Unpublished manuscript, 3 pg. Mehrgarh, March 1984.

Wright, Rita P. and Joan Mishara

ms. Analysis of Kiln Linings and Associated Materials at Lal Shah, Mehrgarh. Unpublished manuscript, 6 pg + 5 figures & 4 XRD printouts. June 1984.

THE JINGDEZHEN[1] EGG-SHAPED KILN

J. Q. Hu, 16125 Howard Landing Dr., Gaithersberg, MD 20878
H. T. Li (1908 -1961), Jingdezhen, Jiangxi, P. R. China

ABSTRACT

One of the most influential kilns in Chinese history is the Jingdezhen egg-shaped wood kiln. The kiln introduced an important technical breakthrough in firing ceramic wares because it used two walls rather than one and utilized the space between them for insulation. The egg-shaped kiln's design, because of the natural flue flow achieved, lends itself to efficient heat transfer. Because it features high-yield percentages and fuel savings, the Jingdezhen egg-shaped kiln influenced Chinese ceramic history from Ming Dynasty to at least the 1930s. The structure and operation of this kiln are discussed along with the thermal behavior.

INTRODUCTION

Jingdezhen is an industrial city in the People's Republic of China, often referred to as the ceramic capital of China. The name of Jingdezhen was given by the emperor of Shang Dynasty bearing the same name--Emperor Jing De--who was honoring the town's brilliant achievements in the ceramic industry at that time. There were many ancient royal ceramic factories in Jingdezhen. These factories manufactured the porcelains for the palace, including dinner wares, decoration wares, and even the porcelain pillow and bed for the emperor's funeral.

While Jingdezhen has a magnificent history of porcelain manufacturing, it still plays an important role in Chinese ceramic industry. It is also well known for the Jingdezhen egg-shaped wood kiln, notable for its energy-saving design, temperature uniformity, and high-yield percentage performance. According to recovered documents, the Jingdezhen egg-shaped kiln originated near the end of the Ming Dynasty (1368-1644) and the beginning of the Qing Dynasty (1644-1911) of China. The Jingdezhen kiln contributed to the most magnificent development in Jingdezhen ceramic history during 1662-1779. Kyozo Suzuki [1955] from Japan observed that designs of both the Newcastle kiln in Britain and the Kasseler-Ofen of Germany during the 18th century were influenced by the Jingdezhen egg-shaped kiln. This shows that the Jingdezhen egg-shaped kiln not only played an important

[1] Zhen means town in Chinese. Jingdezhen was a small town named Jingde.

role in the ceramic history in China, but also had a great influence on European ceramic kiln development.

This paper introduces the structure and operation of the Jingdezhen egg-shaped kiln in two sections. In the first section, the body configuration of the egg-shaped kiln is presented along with the structure of the roof and the wall. The second section describes the operation of the kiln. Information was obtained primarily from *A General Survey of Jingdezhen Industry,* written by Hao-Ting Li in the 1930s. This book discusses Jingdezhen ceramic factories in 1930s, including the categories of ceramic industry, porcelain processing, and firing. The structure and operation of Jingdezhen egg-shaped kiln is also discussed in detail in Hao-Ting Li's book.

The Jingdezhen kiln is a traditional periodical kiln of the Jingdezhen ceramic district. It uses pine wood as fuel. From a distance, the kiln looks like a half-egg laying on the ground; hence, the name of the kiln. Due to the horizontal travel of the flame through the kiln, it falls into the category of horizontal flame kilns (flat flame kiln). The kiln performs outstandingly for two reasons: (1) the structure allows for firing temperatures up to 1350 °C with a relatively stable and uniform temperature profile; and (2) the effective loading method, which is used to load products of different firing temperatures in one kiln at the corresponding temperature zones within the kiln. This enhances the efficiency of the kiln and also provides the flexibility for firing different products. In addition, the ease of controlling the atmosphere within the kiln ensures a high yield percentage performance.

The excellent performance of the kiln is due to its structural advantages. The Jingdezhen egg-shaped kiln introduced an important technological breakthrough in firing ceramic wares, by using two walls and the space between them to improve the insulation of the wall. The egg-shaped construction tailored the velocity of flue flow, enhancing the uniformity of the temperature profile by an optimized heat transfer mechanism. Furthermore, the thermal behavior of the kiln is discussed in this paper in relation to its structure from a modern heat-transfer standpoint.

KILN STRUCTURE

The Jingdezhen egg-shaped kiln uses large pine logs or smaller branches as combustion fuel. Two types of the Jingdezhen kiln, the branch kiln and the log kiln, have the same structure but are of different sizes. The branch kiln is smaller than wood kiln, and the firing temperature of a branch kiln is lower than that of the log kiln, which can reach 1350 °C.

Generally speaking, the Jingdezhen egg-shaped kiln includes two parts: the kiln house and the kiln body. The Jingdezhen kiln is constructed in a kiln house, which protected the kiln from rain and wind and also provided a place for fuel storage and loading processes. Figure 1 gives the schematic of the kiln house arrangement. The kiln body is located at the rear part of the house; the chimney stands out from the roof. In front of the kiln is a large space for loading and sagger storage which is surrounded by an elevated platform along the wall for storing fuel. The efficient arrangement of space makes it convenient for people to load and unload large quantities of the kiln products within the room. The platform design allows the kiln worker to get the fuel easily while keeping the wood dry.

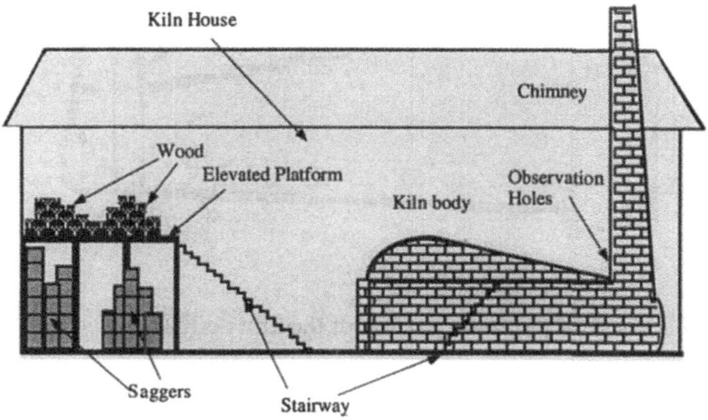

Figure 1. Schematic of the kiln house arrangement.

The Body Configuration

The kiln body consists of six parts, as shown in Figure 2: the door (A), the hearth (B), the ash pit (C), the firing chamber (D), the chimney outthrust (E), and the chimney (F). The hearth of the kiln is located at the front, therefore, it is front fired. Hot gases travel through the kiln and are drawn out of the kiln by the chimney. According to this firing method, the temperature of the gases decreases while they travel toward the chimney; hence, the radiation heat transfer reduces during the process. The heat transfer process is modified by enhancing the convection heat through an increased mean velocity of the gas flow. The control of the gas flow velocity is achieved by tailoring the frontal cross-sectional area of the kiln. From the side view of the kiln (Figure 2), the kiln's roof slopes from front to rear, so that the front part of the chamber roof is higher than the rear. The floor of the kiln tilts up from the front to the rear so that the front floor is lower than that of the rear. Figure 3, the plan view, shows that the width of the kiln chamber decreases across the space from the front to the rear. These features shape a decreasing cross-sectional area while moving toward the end of the kiln. Since the amount of gases passing through each cross-section is the same, the decreased area provides increased velocity, thus enhancing the convection heat. Therefore, temperature uniformity is achieved.

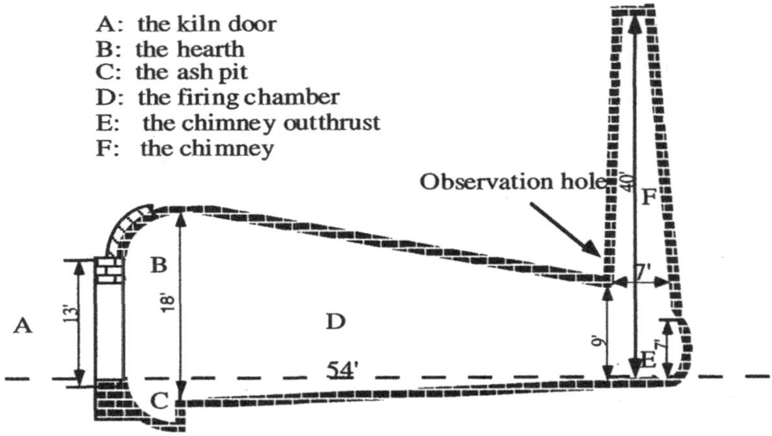

Figure 2. Side view of the configuration of the kiln body.

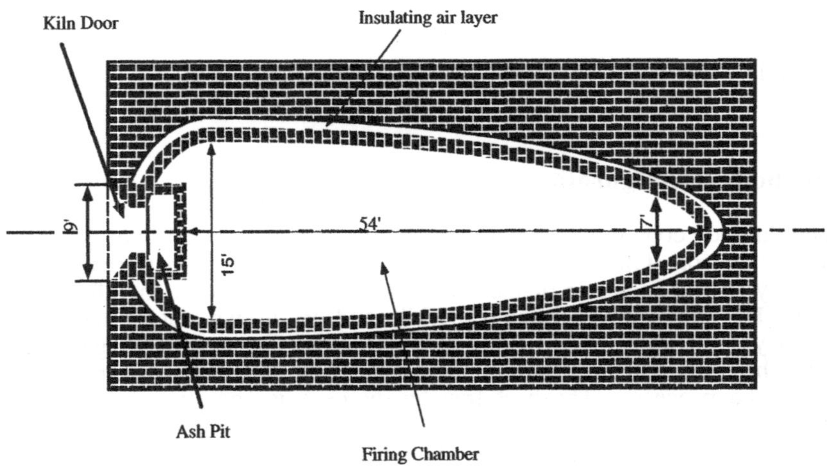

Figure 3. Plan view of the Jingdezhen kiln.

The door of the Jingdezhen kiln is a loose structure composed of refractory bricks and waste saggers. This structure is shown in Figure 4. The lower part of the door is an open pillared stand, which allows the air to go through and the fuel ash to be removed. Above the stands, there is a window for ignition which normally kept sealed. In the middle of the door is a window about one and a half feet square through which wood is thrown into the hearth. About one foot above the window are two holes made of waste saggers[2]. These two holes and another two on the chimney are called firing observation holes. The observation holes on the door play a critical role in determining the firing temperature during the process, which will be discussed later on in this paper.

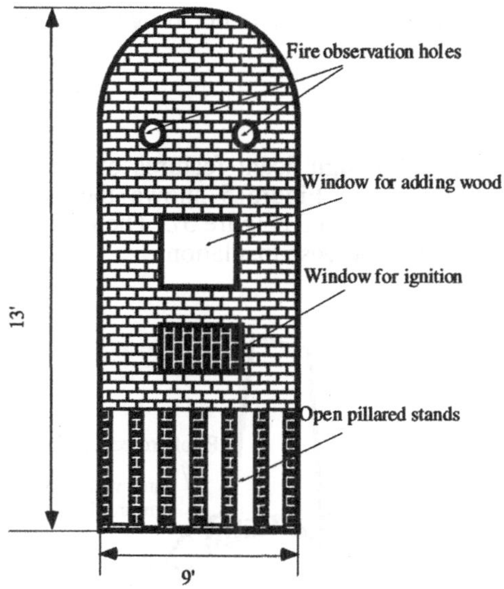

Figure 4. The door structure.

Behind the door is the hearth of the kiln, where the fuel combustion takes place. Brick pillars connected with the door are stacked up from the ash pit, forming a base for the firing grid, while allowing the ash to be removed easily from the kiln. The firing grid, which holds the wood to be fired and allows the ashes to fall through, is constructed from the base using saggers and bricks.

The firing chamber and the hearth form a continuous body. The firing temperature at the front part of the firing chamber is high due to that it is near the hearth, while at the rear, it is lower. Therefore, fine porcelain and colored glaze products, which need to be fired at high temperatures, are normally loaded in the front part of the firing chamber. The middle of the chamber loads relatively low firing temperature

[2] Sagger is refractory container used for firing porcelain.

products. The end of the firing chamber connected with the chimney outthrust has an even lower temperature and is used to fire bricks, roof tiles, and saggers. It also functions to modify the pressure and momentum of the hot flue, in order to protect the chimney from the direct attack by the gases. Another function of the chimney outthrust is to support the weight of the huge chimney.

The kiln chimney is graduated and cone-shaped with an opening at the top. The cross-sections of the chimney are also egg-shaped. Two additional observation holes are located on the front side of the chimney. Unlike the other two observation holes which are normally used to determine the firing temperature, this set of observation holes plays an important role in deciding the firing time.

The Wall Structure

The wall structure is one of the significant features of the Jingdezhen kiln which provides temperature uniformity. Figure 5 shows the structure in detail. The horizontal cross-section is egg-shaped (Figure 3). The kiln makes use of two walls and the space between them to increase insulation.

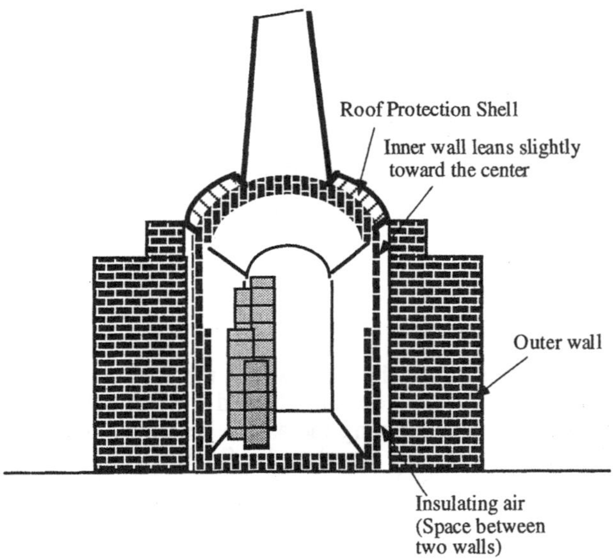

Figure 5. The frontal cross-section of the kiln.

The inner wall is constructed with refractory bricks. It is independent from the outer wall and supports the weight of the roof. Since the inner wall experiences high temperatures, it normally has a shorter life than the outer wall. The unique structure of the wall makes it convenient for kiln repair. The inner wall is about seven to eight feet high and ten inches thick. The space between the two walls is about three inches. Because air has low thermal conductivity, the layer of insulating air dramatically decreased the heat loss from the wall. The good insulation of the wall ensures the temperature uniformity within the kiln. During the firing process, the inner wall expands due to the constantly increasing temperature. The room between the two walls prevents the expansion from damaging the wall. The inner wall leans slightly toward the center of the kiln in order to resist the horizontal force caused by the roof thermal expansion.

The outer wall of the kiln is constructed around the inner wall and has two functions: (1) to insulate and protect the inner wall and (2) to provide a working platform and stairs to access the firing observation hole, which is located at the chimney near the conjunction between the chimney and the kiln body.

The Roof Structure

The roof of the kiln is constructed on the base of the inner wall by the same refractory bricks. It has an egg-shaped horizontal cross-section along with an half circle frontal cross-section. The longitudinal cross-section of the roof has a special curve, shown in Figure 2. The height of the roof decreases from the front to the rear. Coupled with the decreased width of the kiln, the decreasing height contribute to a decreasing cross-section.

In order to insulate the roof, waste saggers were stacked up against the roof as a protecting shell. The air within saggers act as an insulator to prevent the inner roof from exposure to the environment air. This shell also protects the kiln roof from things possibly falling upon it directly.

The Kiln Bottom

As shown in Figure 2, the bottom of the kiln is a slope with the front lower than the rear. This feature as it combined with curve of the roof and the structure of the wall, of which the width decreases as it moves toward the rear, shapes a decreasing cross-section and achieves a flue flow having an increasing gas flow rate toward the rear.

OPERATIONS

The operations of the kiln include loading, unloading, and firing are all conducted by human labor. The firing temperature and firing time are decided empirically by the human eye due to the lack of equipment for precisely measuring the temperature. The procedures are described below.

Loading

One advantage of the egg-shaped kiln is that it can fire different products with different firing temperatures simultaneously through the special loading process. The green products are primarily loaded into saggers, as shown in Figure 6, and stacked up as shown in Figure 7. According to the longitudinal temperature profile, the temperature decreases while moving toward the end of the kiln. Therefore, the lower temperature products such as bricks, saggers, and roof tiles are loaded first at the back of the kiln and the space under the chimney, the chimney outthrust. The porcelains with higher firing temperature are then loaded into the middle part of the chamber. At last, porcelains with the highest firing temperature, such as fine porcelain and products with colored glaze, are loaded at the front area of the firing chamber, where is the closest to the hearth that the temperatures can reach 1350 °C.

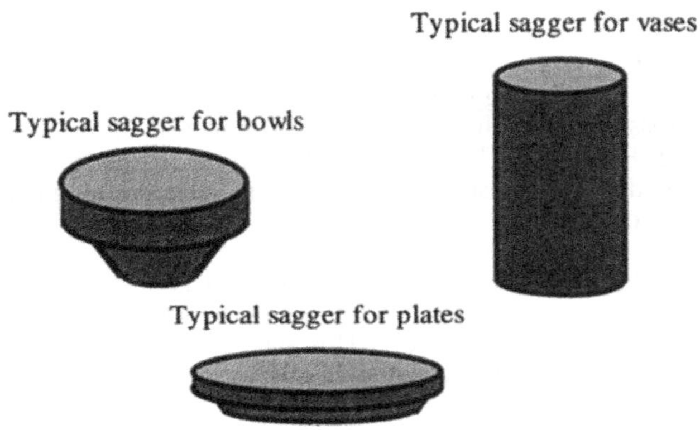

Figure 6. Typical saggers shapes for loading bowls, plates, and vases.

The sagger stacks are set on sagger stands, which allow the gas to flow through. The arrangement of the sagger stacks is shown in Figure 7. Each row is offset from the row in front of and behind it. The space between every two saggers is about two to three inches, and is inserted with broken saggers to make the sagger stacks fit tightly against each other in order to prevent the momentum of the hot gas flow from blowing them down.

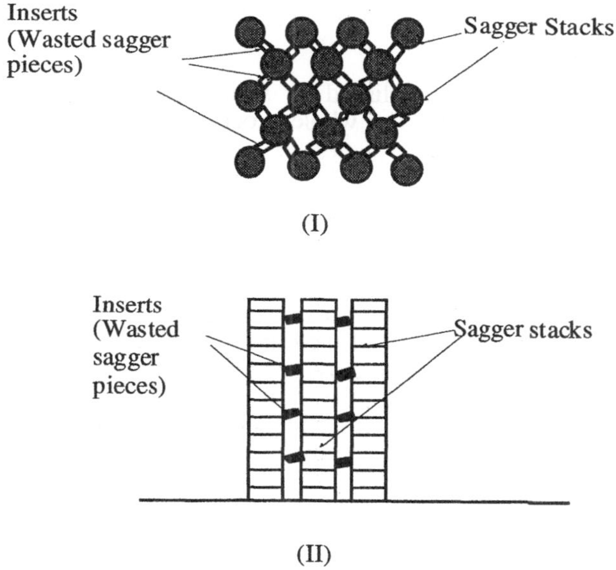

Figure 7. Schematic of the sagger loading arrangement.

Firing

When the kiln is fully loaded, the door is constructed. Then, the branch on the firing grid is lighted first through the igniting window. After ignition, the igniting window is sealed with bricks. Wood is constantly thrown through the firing window on the door until a certain firing time is reached.

There is no equipment for measuring temperature at that time; the whole firing process is entirely determined by one person called the "Kiln-Eye", who can tells the temperature and atmosphere by naked eye observations. The "Kiln-Eye" plays a key role during the firing process. He controls the whole firing process. The amount of fuel being added each time, the time to add fuel, and the time to stop firing are all decided by the "Kiln-Eye". The saggers used to make firing observation holes on the door provides some information which helps the "Kiln-Eye" to determine the firing process. The atmosphere is achieved by controlling the amount of fuel added the opening the pillared stands at the door.

When the saggers used as observing holes change the color from red to white, the temperature within the kiln reaches above 1300 °C. Then, the "Kiln-Eye" will climb up to the roof near the chimney, remove the lid of the firing observation hole in the chimney, and observe the condition inside the kiln through these holes. When the saggers are totally white and the atmosphere is clear, the "Kiln-Eye" spits saliva into the observing hole. If the saliva vaporizes into a white vapor and rises up into the chimney immediately, it is the time to stop firing. Otherwise, the firing process continues.

The cooling procedure is accomplished by simply pulling down the door, and letting the air get into the kiln. The whole process takes about a day and a half before the kiln can be unloaded.

SUMMARY

The Jingdezhen egg-shaped kiln uses two walls and the space between them to enhance the insulation of the walls. This structure provides temperature uniformity and increases the energy efficiency of the kiln. The other method which achieves the same goal is to use the decreased cross-section from front to rear to increase the gas flow rate, thereby increasing the heat convection. The increased heat convection compensates for the decreased radiation heat transfer while moving from the front of the kiln to the rear. The overall heat transfer, therefore, has been modified.

The utilization of two set of observation holes solves the difficulties of determining temperatures and firing times. Because the rear part of the kiln heats up more slowly than the front, the time difference between them provides the information to determine the firing time.

Due to the fact that the qualities of ceramics not only depend on the properties of the raw material, but also rely on the firing equipment, Jingdezhen egg-shaped kiln had facilitated the ceramic industry of Jingdezhen to be developed significantly during the end of the Ming and begin of the Qing dynasties.

Acknowledgment - This paper is presented in the memory of my grandfather Hao Ting Li. As a ceramic engineer, he had devoted his whole life in the effort of developing the Jingdezhen ceramic industry. I would like to dedicate this paper to my family, who gave me strong support on pursuing my career goal to become a good ceramic engineer. Special thanks are extended to Dr. Alastair Cormack, Dr. George Onoda, and Dr. Vernon Burdick for their support. I am in gratitude of my advisor, Dr. Linda Jones for her support, understanding, and encouraging. The help from Dr. Sylvia Bryant for spending time to review my paper is gratefully acknowledged. I would like also thank my friends for their support and presence in my talk. Special thank to Aaron Blanchet, Brett Thorsland, David Andrews, Jack

Debolt, and James Doran for their generously taking time to help me prepare the presentation.

I also thank Dr. W. David Kingery and Dr. Prudence Rice for organizing such a wonderful section and providing me the chance to present this paper.

REFERENCES

Li, H. T.
1930 *A General Survey of Jingdezhen Industry,* Zhongzhen Publication Bureau, Nanjing, Zhenjiang Province, China.

Liu, Z., Zhen, N. Z., and Hu ,Y. Z.
1984 "The Study of Configuration and Construction Technique of Jingdezhen Kiln" Journal of the Ceramic Institute of Jingdezhen 5[2], Pp. 17.

TECHNOLOGY AND ORGANIZATION OF ANASAZI TRENCH KILNS

Eric Blinman, Museum of New Mexico, Office of Archaeological Studies, PO Box 2087, Santa Fe, NM, 87504-2087

Clint Swink, Swink Art, 688 Raven Ridge, Bayfield, CO 81122

ABSTRACT

Trench kilns were used in the production of Mesa Verde region Anasazi black-on-white pottery between at least A.D. 900 and 1250. Although first suspected in the 1940s, the kiln function was only formally recognized in the 1970s and 1980s. Many examples are in isolated settings, they often occur in clusters, some are extremely long, and none is associated with either gray- or red-ware vessel production. After 10 years of excavation and experimental research, we are close to understanding how these kilns were used to produce white-ware vessels. In addition, implications for production organization are tantalizingly complex. The emerging picture of the firing technology suggests that the large kilns represent collaborations of several potters rather simple conceptions of large scale production by a skilled individual.

INTRODUCTION

In the Southwestern United States, the Anasazi culture occupied an area that is centered today on the Four Corners, where the borders of Colorado, New Mexico, Arizona, and Utah meet (Figure 1). Although pottery is a ubiquitous aspect of Anasazi archaeology, progress in understanding ceramic production technology and organization has been relatively recent, despite a tremendous foundation laid by the work of Anna O. Shepard (1939, 1965; Bishop and Lange 1991). The major impediment has been a lack of archaeological information on Anasazi firing techniques. Without such information, archaeologists have tended to rely on ethnographic descriptions of pottery production (such as Guthe 1925)

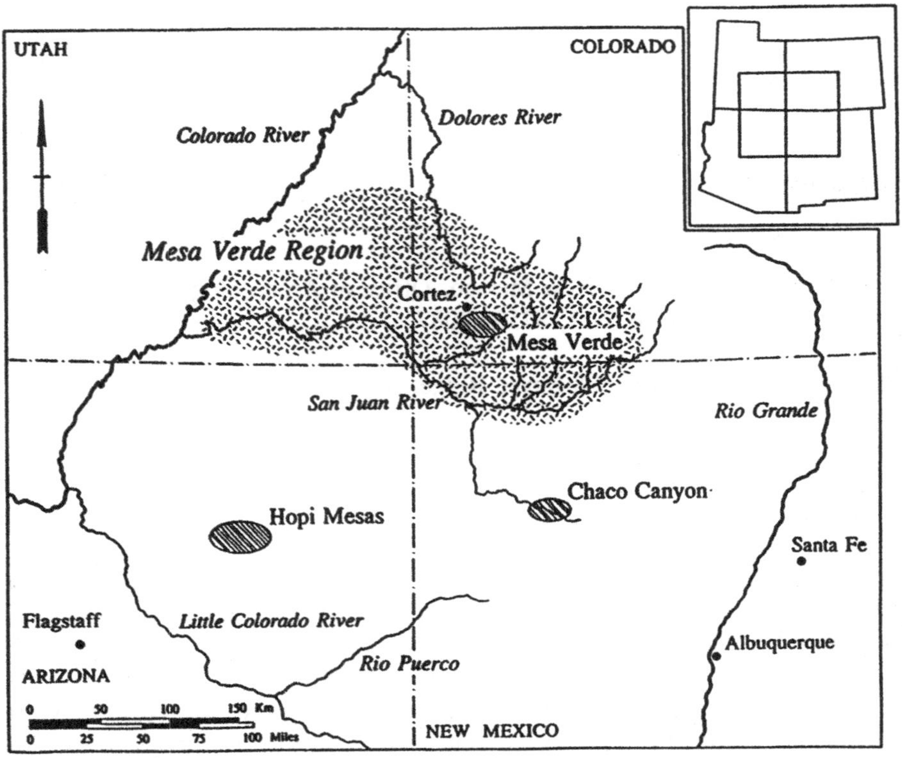

Figure 1. The Mesa Verde or Northern San Juan region of the Anasazi Culture area.

for interpretive models of production technology and organization. This situation changed in the early 1970s with the recognition of the first archaeological firing features or kilns (Helm 1973) and with the recognition of their abundance in the 1980s (Fuller 1984).

Consisting of shallow stone-lined trenches (Figure 2), these kilns were clearly different in design and scale from ethnographic and historic firing features. Interpretive models based on the firing techniques of Native American potters were no longer applicable, and new models were needed through theoretical and experimental research. In 1991, Crow Canyon Archaeological Center in Cortez, Colorado, hosted a conference of archaeologists and potters, including Native Americans. This began an informal but intense collaboration, the goal of which has been the experimental study of ancient Anasazi firing techniques. Since then, results of individual experimentation have been shared at annual Kiln

Figure 2. Trench kiln under excavation by La Plata Archaeological Consultants in 1982.

Conferences. Progress in understanding firing principles has been dramatic, both answering and raising questions concerning the prehistoric technology.

ANASAZI TRENCH KILNS

Although hundreds of possible Anasazi trench kilns now have been identified by archaeological surveys, only about 30 have been excavated. The vast majority of the kilns, and all of the excavated examples, have been located in the Mesa Verde or Northern San Juan region (see Figure 1). Other kiln types have been identified in other regions (Post and Lakatos 1995), and although the area of trench kiln distribution may enlarge with additional archaeological research, the trench kiln type appears to be a northern style.

Descriptions of 19 excavated trench kilns have been summarized by David Purcell (1993), 9 kilns have been reported in draft form from recent excavations at Mesa Verde National Park (Brisbin 1996), and several other kiln features have been described individually (Honeycutt and Fetterman 1994) or are still being studied. The excavated kilns range in age from as early as A.D. 900 through the Anasazi abandonment of the region prior to A.D. 1300. Most of those that can be firmly dated fall within the Pueblo III period (the latter third of this range), but many cannot be dated with confidence. Given the similar qualities of pottery for the entire Anasazi time span (A.D. 600-1300) it is probable that analogous if not identical features will be discovered for the pre-A.D. 900 period.

Trench characteristics show strong regularities but some variability as well. Most features are rectangular in plan, although some depart toward irregular or oval outlines. Lengths range from 1.5 to 8.5 m (Figure 3). Width is the most uniform dimension, ranging from 0.8 to 2.0 m, while depth is variable, ranging from 10 to 65 cm. With two exceptions, there is a strong correlation between increasing width and increasing depth (C. Swink 1996a). Although there are some suggestions of increasing maximum length through time, the sample of securely dated features is too small for confident interpretation, and shape and size variability is considerable even within the late examples. Also, although all kilns have a constructed length, some have clear evidence of later subdivision into smaller firing chambers, and some have areas or shelves of shallower depth, perhaps to accommodate different vessel sizes.

Stratigraphic contents of the kilns can be generalized with little qualification (Figure 4). In nearly all cases, upright slabs, usually sandstone, line the sides and ends of a shallow trench. The trench base is covered with a layer of charcoal-rich earth, often including large fragments of partially consumed fuel. On this layer is a nearly continuous but often irregular pavement of well-burned and blackened stones that fills the entire plan of the trench. Pottery, if present, overlies the stones in a matrix of charcoal and earth. Usually, a final upper layer consists of earth with distinctly less charcoal. Charcoal is dominated by woody species such as piñon pine and juniper, with only occasional preserved evidence of shrubby fuels. Although vegetation zones with scrub oak are present in the Northern San Juan region, no kilns have as yet been found in those zones, and oak has not been identified as a significant fuel component. Extremely high corn pollen content suggests that corn stalks were a secondary fuel in at least one case (Honeycutt and Fetterman 1994:22.2-7).

Pottery is surprisingly scarce in these features and falls into three categories. Usable vessels are extremely rare and are represented only by small mugs and a ladle that were found on the pavement, perhaps items that were overlooked on emptying the kilns after the last use. Slightly more common are miniature pinch pots or disks of clay that may have been potters' test pieces or children's vessels. These either were not retrieved or were discarded in the kiln after retrieval and examination. Broken pottery, partial vessels, or bloated fragments are more

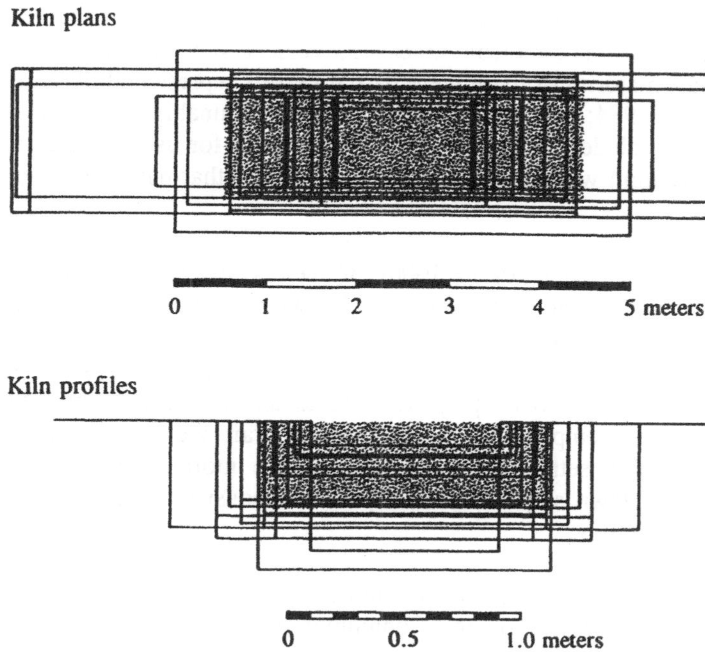

Figure 3. Schematic diagrams of kiln dimensions: length and width, and width and depth. Shapes have been simplified as rectangles, and shaded areas represent mean dimensions. Data are from Purcell (1993:Table 2).

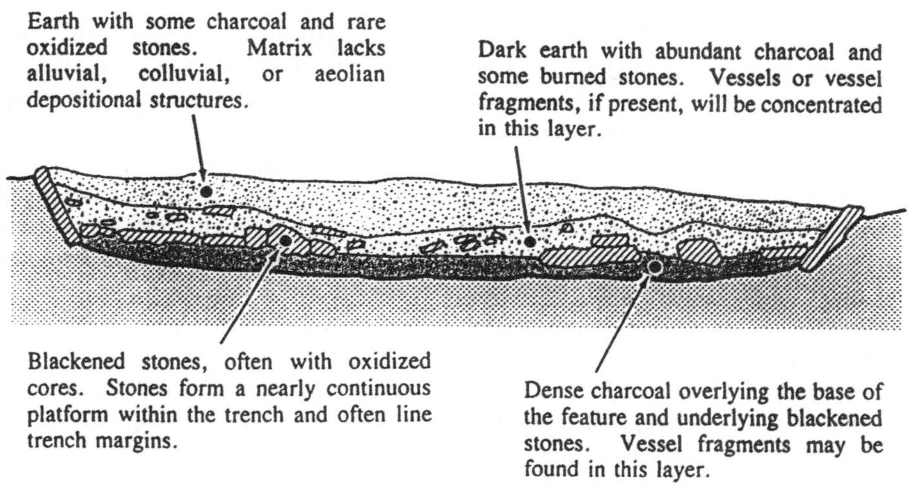

Figure 4. Representative stratigraphy of Anasazi trench kilns.

common, but significant amounts are present in only a minority of the kilns. In the most extreme case, sherds from at least 35 vessels were identified in a single kiln, but the vast majority of these were partial vessels represented by only a few fragments (Fuller 1984; C. Swink 1996b). These presumably are dominated by wasters but may include some cover sherds left behind for use in future firings. Ceramics of any kind were absent in 7 of the 28 kilns that have been described in detail.

Kilns can be located from 10s to 1000s of meters away from contemporary residential sites, and in one case the nearest known residence is more than 5 km away. Slight slopes are favored but not exclusively so, and some kilns are placed in shallow drainage bottoms or at the convergence of small and larger drainages. Feature orientation tends to be perpendicular to dip direction, suggesting that topography was not employed to encourage a draft along the kiln length. However, geographic settings are in places where morning and evening air movement up and down slope may have provided consistent light side breezes during parts of each day.

A singular locational aspect is that kilns can occur in clusters of up to six features within 10s of meters of each other. Contemporaneity is difficult to assess with the available dating techniques, and in only a few cases are kilns so close as to preclude simultaneous use.

FIRING REGIME

Without appropriate ethnographic analogues, experimental replication has been necessary to develop models of kiln function. Guidelines for the replication experiments have been the characteristics of the trench kilns, the raw materials and fuels used by the prehistoric potters, and the appearance of Anasazi pottery. These guidelines have insured a degree of authenticity, but we acknowledge that successful replication need not mean that prehistoric practices have been duplicated exactly. Experimental research from 1991 through 1996 has closely approximated most aspects of trench kiln appearance and stratigraphy, as well as the qualities and appearance of Anasazi black-on-white vessels. The greatest progress occurred through close collaboration between Clint Swink, an artist and potter, and Mesa Verde National Park archaeologists who were engaged in a series of kiln excavations (C. Swink 1996a). Although minor discrepancies in the current firing model suggest that some aspects of the Anasazi firing regime are still imperfectly understood, the structure of the current model appears to be robust (Figure 5) (C. Swink 1993).

Preliminary fire -- A fire is constructed within the kiln and allowed to burn to produce a layer of charcoal across the trench base.

Secondary fire -- The stone platform, vessels, and cover sherds are placed on the charcoal layer. Fuel is ignited over the setting.

Quenching -- After the fuel burns to charcoal and collapses, the setting is covered with a smothering layer of earth and allowed to cool.

Figure 5. Firing sequence model based on trench kiln replication experiments.

After trench construction, a fire of coarse fuel is set within the feature. When the fuel is reduced to a layer of coals, a stone pavement is constructed on the coal bed. Vessels are placed in a single layer on this pavement, and if cover sherds are used they are arranged at this time. As the vessels heat up from the underlying coals, a dome or layer of wood fuel is built above the vessels, supported by the edges of the trench and with air space between the vessels and the fuel. Controlled ignition of the overlying fuel radiates heat downward over the vessels and encourages drafts that raise the temperature of the underlying coals (Figure 6). A reduction atmosphere is maintained in the early stages of firing by volatile gases released by the fuel and by any carbonaceous material in the clay. During this phase, some blackening or sooting of vessel surfaces can occur. As combustible gases are exhausted and charcoal alone provides both upper and lower heat, a slight oxidation atmosphere prevails, removing soot. By the time the overlying fuel collapses on the vessels, the effects of temperature are nearly complete. After short soaking near peak temperature, the firing is quenched with a layer of earth, isolating both the fuel and vessels from additional oxygen. Temperature declines steadily, and within 24 hours the vessels are usually cool enough to remove without the risk of additional oxidation that would alter vessel appearance.

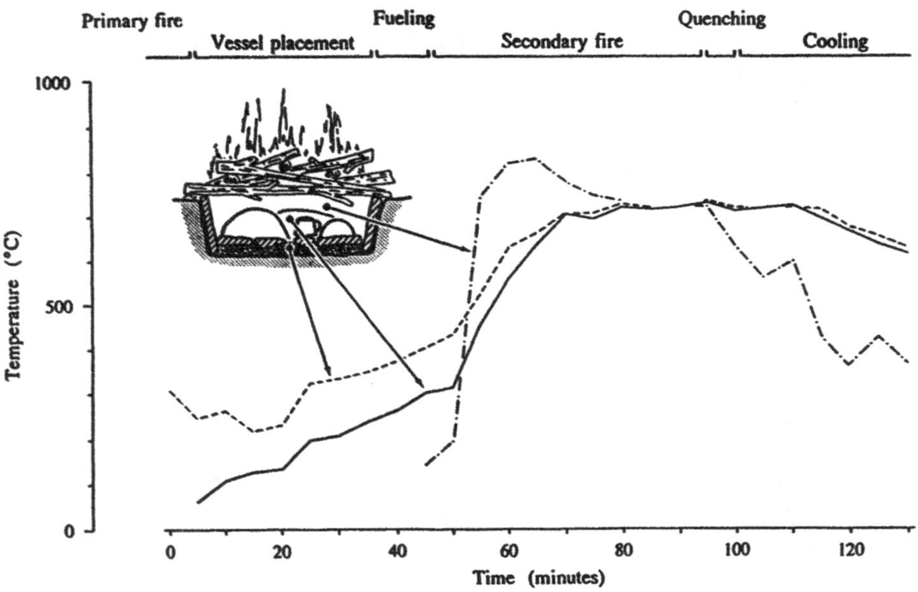

Figure 6. Experimental firing temperatures. This firing did not reach optimal temperatures (800-900° C), but it demonstrates the shape and timing of a typical firing sequence. Thermocouples were placed at different points within the setting.

Fuel selection and manipulation are key to achieving adequate temperatures and proper atmosphere control. Heat rise that is too rapid can cause vessels to shatter from differential thermal expansion and steam evolution. Over-fueling can slow combustion, lower temperatures, and subject vessels to too much reduction, resulting in soft and blackened vessels. Underfueling can also result in lower temperatures and can produce an inadequate reduction phase, resulting in soft and occasionally oxidized vessels. Impatience in waiting for the exhaustion of volatile fuel gases either before vessel loading or before quenching can cause smudging. Finally, if the late oxidation phase lasts too long before quenching or is too intense, organic-painted designs can be lightened or erased, and iron compounds in mineral paint or the clay can be oxidized. This over-oxidation results in weak designs and yellow or pink background colors, a degradation of vessel appearance that is seen on only a minority of Anasazi pottery.

Laboratory experiments currently being conducted by Lawrence Sitney at Los Alamos National Laboratory have demonstrated the importance of the reduction phase in Anasazi firing (Sitney et al. 1995). Under reducing conditions, the sintering curve for the shale clays used by Anasazi potters is effectively enhanced. Under reduction conditions, a well-fired ware can be produced at 800° C, while a temperature of closer to 1000° C would be required to produce the same degree of sintering under oxidizing conditions. In experimental trench kiln firings, temperatures of 800-900° C produce wares comparable in quality to the majority of Anasazi pottery. This staged atmosphere control of the Anasazi firing regime appears to be essential to vessel strength as well as having the secondary effect of producing the characteristic white or gray background for the painted designs.

Several points continue to be investigated through firing experiments. These include the role of cover sherds and minimization of fuel volume. The question of cover sherds is particularly intriguing. Only a few archaeological kilns contain sherds that could have been used as covers, and even these may have been wasters. Even in the large isolated kilns where retrieval of cover sherds for reuse elsewhere would have been less likely, there are insufficient sherds left in the kilns to have protected any significant portion of the setting. However, during experimental firings cover sherds effectively protect vessels from heat shock during the ignition of the overlying fuel, they can enhance temperature through draft control immediately after collapse of the upper fuel layer, and they provide vessels with some protection from the physical and atmospheric effects of both overlying fuel collapse and the subsequent earth covering. If cover sherds were used routinely, Anasazi potters must have valued them sufficiently to retrieve them after the final use of many of these trench kilns, even those at some distance from residential sites. Alternatively, the replicated firing regime may be an inaccurate characterization of the ignition and fuel collapse phases, with Anasazi potters following different procedures that minimized the risks that the experimental firings now mediate with the cover sherds.

Fuel consumption during initial experimental firings was relatively great, and with increasing experience has been reduced to an average of about 6.5 kg per vessel or about 100 kg per square meter of kiln area (C. Swink 1996a). Although current experimental fuel use patterns have been an improvement, the figure is still high. If 100 kg per meter is a reasonable approximation, fuel demand would have been a strong impetus toward moving firing locations to areas of abundant fuel availability. However, much of the heat generated by combustion is lost to upward radiation, and continuing experimentation with fuel selection and arrangement suggests that additional improvements in efficiency can be achieved. These improvements appear to hinge on fuel quality (shape and size) and arrangement. Fuel quality may prove to be as important in Anasazi kiln location decisions as fuel abundance.

IMPLICATIONS FOR PRODUCTION ORGANIZATION

Knowledge gained through replication experiments has gone beyond technical aspects of the firing regime and extends to questions of production organization. It is tempting to assume large scale production specialization based on the sheer size of the largest kilns and the tendency for facility investment to correlate with degree of specialization (Arnold 1985:219). However, the situation is far more complex.

Size does mean large numbers of vessels in each firing. Experimental firings can easily accommodate 15 vessels per square meter of kiln area, implying a range of about 25 to 200 vessels per firing for the prehistoric kilns. Estimates of the number of vessels in Anasazi households can range up to 30, including a full range of vessel forms, half or more of which would be decorated wares (Blinman 1988:152-156). Despite this large inventory, ethnographic and archaeological data on vessel breakage rates suggest that replacement of as few as five vessels per year would be a reasonable estimate (Kohler and Blinman 1987), perhaps two or three of which would be white wares in a given year. A single firing of even the smallest kiln would produce enough white-ware vessels for the annual needs of several households, and the largest kilns could produce enough replacement vessels for as many as 80 households.

Firing frequency is also an important consideration in estimating kiln production volume. This variable is impossible to determine with presently available temporal control, but kilns were clearly used repeatedly. This is suggested by the depth of some kiln oxidation rinds, by the degree of fragmentation of some of the pavement rocks, and by episodes of kiln remodeling. However, whether firings were scheduled once or several times each year is unknown and perhaps unknowable.

Repeated use is also supported by the use of slab linings for the sides and ends of most trenches. Initial interpretation of linings included the possibility that the slabs served to reflect heat into the firing feature, enhancing the firing temperature. Experimental results suggest that fuel manipulation plays a far more important role in heat generation, and there is no significant decrease in firing temperature associated with the use of a simple earthen trench as is seen in some kilns. It is far more likely that the stone linings were designed for feature permanence. The stones would keep the sides of the kiln from collapsing during and between uses, and the lining would allow easy cleanout for the next use. This pattern of reuse is consistent with the occasional remodeling or subdivision of kilns, in which kiln size is temporarily reduced to accommodate a smaller vessel load.

The presence of multiple kilns within meters of each other also affects our interpretation of production volume. This proximity can be interpreted as simultaneous use by multiple potters with correspondingly high production volume. However, simultaneous versus sequential use of adjacent kilns is nearly impossible to differentiate archaeologically. Native American potters have suggested an alternative interpretation. Traditional potters will frequently abandon a kiln if a firing fails. Moving the firing location a few meters away is often sufficient, and the potter may or may not reuse any of the construction materials from the failed kiln. In one case where five kilns were excavated at a single site, all contained significant quantities of wasters. Unfortunately, this content could be consistent both with a simultaneous use model and with sequential use following failures.

Kiln locations provide additional points to consider in evaluating production organization. The isolated nature of many kiln locations has prompted speculation concerning fuel availability. Anasazi settlements were heavy consumers of fuel for domestic purposes, often depleting local supplies through the lifespan of villages (Kohler and Matthews 1988). Patterns of domestic fuel use appear to deplete woody species first, followed by depletion of shrubby fuels. Since the archaeological evidence and experimental firing regime suggest reliance on charcoal-producing fuels, pottery firing would require substantial quantities of high-quality dry wood. Where local supplies were depleted or scarce near residential locations, moving the firing to the fuel source may have been more efficient than moving the fuel to the firing.

However, experiments with pottery workshop participants have shown that moving unfired vessels as much as several kilometers to a kiln site poses its own problems. Unfired pottery is weak, and painted designs are tacky and can be smudged by physical contact. This fragility means that vessel transport is not without risk. Pottery production at the kiln site is possible, but there are as yet no clear suggestions of any significant residence time at the outlying kiln locations. Accessory hearths are rare, and there has been no evidence of permanent or ephemeral structures in the vicinity of the outlying kilns. Nonperishable artifactual evidence of pottery production is rare, even on sites with definite evidence of production (Blinman 1988:76-97), and although the lack of pottery production tools at kiln sites is consistent with production elsewhere, it is a weak argument.

Clay source distributions can influence the location of production facilities in some cases, including firing facilities. Informal surveys in the vicinity of some trench kilns have located clay sources within 100s of meters. However, clay sources are so common amid the dissected topography of the Colorado Plateau that few areas of the northern Anasazi landscape are outside of the resource catchments characteristic of ethnographic potters (Arnold 1985:35-57). We do not believe that clay sources are more strongly associated with kiln locations than with residential locations.

Seasonal weather patterns would have further influenced the organization of production and outlying kiln use, although the influence of these patterns would have been much less for firing features that were proximate to the potters' residences. Freezing evening temperatures can occur from September through May. Although cold does not affect the actual firing (successful experimental firings have been conducted in subfreezing temperatures), if freshly formed vessels are left unprotected and freeze while awaiting firing, they are at risk of noticeable strength reduction (T. Swink 1996). Another constraint is imposed by afternoon thunderstorms that are common from July through September. These rains need not interfere with the firing since the process can be carried through the quenching stage in the dry morning hours. However, vessels cannot be stockpiled and left unprotected at a distant firing location during the monsoon season. This would also affect production at the outlying kiln site, since many days would be required for the forming, decoration, and drying of hundreds of vessels.

Where each kiln reflects a single large-volume firing, settings do not appear to have been limited to the products of a single potter. Potential collaboration of multiple potters is difficult to evaluate, since there is no way of determining conclusively whether kiln contents (wasters or cover sherds) are cumulative remains from multiple firings or whether the associated ceramics are the consequence of the last firing only. However, since the experimental firing model requires that the trench be cleaned out before each use, it is likely that wasters (as opposed to cover sherds) are dominated by the vessels from the last firing. If this is true, then it is likely that some trench kiln firings were collaborative events. The contents of several kilns have been examined from the perspectives of both archaeologists and artists (Fuller 1984; C. Swink 1996b). Individual kilns have included ceramics with multiple and distinctive paste-slip formulations, and several kilns have included both mineral- and organic-painted vessels. These combinations are unlikely to have been the products of a single potter. Design styles in these cases have also been diverse, although this attribute is more difficult to interpret confidently.

Although the lines of evidence are not conclusive, we believe that the larger firings were collaborations of multiple potters. The argument is strongest for the large outlying kilns, since they are most affected by the risks to unfired pottery of long distance transport and weather exposure. However, we suspect that the collaborative model is applicable to the trench kilns in general.

Fuel efficiency gain in one large versus several smaller firings would encourage collaborations, but there are stronger logistical and technical arguments. Vessel quantities implied by the large kilns are not outside the conceivable range of a potter's annual production, but there is little room in Anasazi dwellings to accommodate the storage of up to 200 vessels leading up to a firing. Transporting vessels would have required considerable numbers of people, arguably stretching the abilities of a single family. Fuel gathering also represents a substantial effort, with conceivably more than a metric ton needed for the largest firings. Fuel manipulation during firing is also critical and would benefit from multiple participants. Fueling a 4-meter demonstration kiln at the 1994 Pecos Conference required the attention of four people during the most active stage of the firing. This labor demand is within the range of a family, but it is not an unskilled task and would benefit from experience. Collaborations would allow the most experienced potter to serve as firing director, having the secondary effect of transmitting knowledge through communities and generations of potters. For the large outlying firing features, the summer solstice would have been an auspicious time for firings, both in terms of weather and in minimizing scheduling conflicts with agricultural tasks.

CONCLUSION

The general consensus of archaeologists in the American Southwest is that Anasazi black-on-white pottery production was specialized at the level of a household industry (Mills and Crown 1995; Wilson and Blinman 1995). Supporting arguments have hinged on a relatively narrow range of clays used in white-ware production (Hegmon et al. 1995), widespread exchange of these wares, a few burials of women with abundant pottery production tools as grave goods (Crotty 1983), and the existence of large specialized kiln features. This production model assumes scattered skilled individuals, with slight regional concentrations or community specialization based on resource availability and quality. Increased specialization through time is suspected, in part on the basis of the many large trench kilns that date to the Pueblo III period, but still within the household industry concept (Wilson and Blinman 1995).

This late increase in apparent specialization may be overstated if the collaborative model suggested by the replication studies is correct. Other variables such as increasing community nucleation in the Pueblo III period (Lipe 1995) and fuel depletion could be contributing to an increase in the size of collaborative groups, with the consequence that the degree of specialization may be less than feature size alone would suggest.

Exchange patterns are partially consistent with this view of less extreme specialization. The detectable volume of total pottery exchange actually decreases in the Pueblo III period (Blinman and Wilson 1993). Communities within the Northern San Juan region are importing fewer vessels from outside the region, and there are fewer indications of within-region vessel distribution. There does seem to be some exportation of Northern San Juan vessels to areas such as Chaco Canyon to the south, but the volume is not extreme and from the Chacoan perspective this is a change in source area rather than necessarily a change in the volume of imports (McKenna and Toll 1991; Toll et al. 1980). The majority of pottery production from the trench kiln appears to be circulating within communities and local areas rather than over larger distances. The trench kilns appear to be indicative of large-scale firing, but not necessarily large-scale specialization.

Pottery production must be understood in terms of both technology and organization, and the prehistoric Anasazi trench kilns are graphic examples of this principle. Strictly archaeological data have been limited in the picture that can be developed of trench kiln use, and an aggressive program of replication experiments has produced valuable complementary information. Knowledge gained through replication should not be considered accurate or absolute, but it does raise important issues and provide new perspectives on the archaeological data. In this case, replication emphasizes the probable role of collaboration among groups of potters in firing white ware vessels.

ACKNOWLEDGMENTS

Margaret Heath and C. Dean Wilson initiated the Kiln Conference at the Crow Canyon Archaeological Center and deserve acknowledgment for the research products that have grown out of that inspired moment. Over the years, the staff of Crow Canyon and conference participants have been essential to the success of the replication experiments, challenging complacency and producing hundreds of sacrificial vessels. Paul Ermigiotti deserves special thanks for conference organization through the years, and ethnobotanists Mark Hovezak, Karen Adams, and Suzanne Edminster have labored amid chaos to quantify fuel use. Clint Swink's workshop participants and family have served as guinea pigs, testing ideas ranging from vessel transportation problems to freezing effects. Larry Sitney has supported both low and high tech research approaches, beginning with temperature measurement of experimental firings and continuing with laboratory experiments in cooperation with the ceramics research group of Los Alamos National Laboratory. C. Dean Wilson and H. Wolcott Toll have directly and indirectly contributed to this research through their depth of knowledge of Anasazi ceramics and their long term concern with issues of production and exchange. Finally, Prudence Rice, Julia Costello, and the other participants in the symposium provided valuable comments and suggestions to improve the form and content of this paper.

REFERENCES

Arnold, Dean E.
1985 *Ceramic Theory and Cultural Process*. Cambridge University Press, New York.

Bishop, Ronald L., and Frederick W. Lange, eds.
1991 *The Ceramic Legacy of Anna O. Shepard*. University Press of Colorado, Boulder.

Blinman, Eric
1988 *The Interpretation of Ceramic Variability: A Case Study from the Dolores Anasazi*. Unpublished Ph.D. dissertation, Department of Anthropology, Washington State University, Pullman.

Blinman, Eric, and C. Dean Wilson
1993 Ceramic Perspectives on Northern Anasazi Exchange. In *The*

American Southwest and Meso America: Systems of Prehistoric Exchange, edited by J.E. Ericson and T.G. Baugh, pp. 65-94. Plenum Press, New York.

Brisbin, Joel M.

1996 Mesa Verde Segment Pottery Firing Kilns. In *Mesa Verde Waterline Replacement Project Phase III Archeological and Historical Studies*, by G.A. Ives, J.M. Brisbin, J.A. Bell, and C.L. Williams. Manuscript on file, Division of Research and Resource Management, Mesa Verde National Park, Colorado.

Crotty, Helen K.

1983 *Honoring the Dead: Anasazi Ceramics from the Rainbow Bridge-Monument Valley Expedition*. University of California, Los Angeles, Museum of Cultural History, Monograph Series 22.

Fuller, Steven L.

1984 *Late Anasazi Pottery Kilns in the Yellowjacket District, Southwestern Colorado*. CASA Papers 4. Complete Archaeological Service Associates, Cortez, Colorado.

Guthe, Carl E.

1925 *Pueblo Pottery Making: A Study at the Village of San Ildefonso*. Published for the Phillips Academy by Yale University Press, New Haven.

Hegmon, Michelle, Winston Hurst, and James R. Allison

1995 Production for Local Consumption and Exchange: Comparisons of Early Red and White Ware Ceramics in the San Juan Region. In *Ceramic Production in the American Southwest*, edited by Barbara J. Mills and Patricia L. Crown, pp. 30-62. University of Arizona Press, Tucson.

Helm, Claudia

1973 The Kiln Site. In *Highway U-95 Archeology: Comb Wash to Grand Flat*, assembled and edited by Gardiner F. Dalley, pp. 209-221. Department of Anthropology, University of Utah. Submitted to Utah State Department of Highways.

Honeycutt, Linda, and Jerry Fetterman, eds.

1994 *Excavations Along the Arkansas Loop Pipeline Corridor, Northwestern New Mexico*. Woods Canyon Archaeological Consultants, Yellow Jacket, Colorado.

Kohler, Timothy A., and Eric Blinman

1987 Solving Mixture Problems in Archaeology: Analysis of Ceramic Materials for Dating and Demographic Reconstruction. *Journal of Anthropological Archaeology* 6:1-28.

Kohler, Timothy A., and Meredith Matthews

1988 Long-Term Anasazi Land Use and Forest Reduction: A Case Study from Southwest Colorado. *American Antiquity* 53:537-564.

Lipe, William D.
1995 The Depopulation of the Northern San Juan: Conditions in the Turbulent 1200s. *Journal of Anthropological Archaeology* 14:143-169.

McKenna, Peter J., and H. Wolcott Toll
1991 Ceramics from 29SJ 633, the Eleventh Hour Site. In *Excavations at 29SJ 633: The Eleventh Hour Site, Chaco Canyon, New Mexico*, edited by Joan Mathien, pp. 139-205. Reports of the Chaco Center, No. 10. US Department of the Interior, National Park Service, Branch of Cultural Research, Santa Fe, New Mexico.

Mills, Barbara J., and Patricia L. Crown
1995 Ceramic Production in the American Southwest: An Introduction. In *Ceramic Production in the American Southwest*, edited by Barbara J. Mills and Patricia L. Crown, pp. 1-29. University of Arizona Press, Tucson.

Post, Stephen S., and Steven A. Lakatos
1995 Santa Fe Black-on-white Pottery Firing Features of the Northern Rio Grande Valley, New Mexico. In *Of Pots and Rocks: Papers in Honor of A. Helene Warren*, edited by M.S. Duran and D.T. Kirkpatrick, pp. 141-154. The Archaeological Society of New Mexico, 21. Albuquerque.

Purcell, David E.
1993 Pottery Kilns of the Northern San Juan Anasazi Tradition. M.A. Thesis, Department of Anthropology, Northern Arizona University.

Shepard, Anna O.
1939 Technology of La Plata Pottery. In *Archaeological Studies in the La Plata District: Southwestern Colorado and Northwestern New Mexico*, by Earl H. Morris, pp. 249-287. Carnegie Institution of Washington Publication 519.

1965 *Ceramics for the Archaeologist*. Carnegie Institution of Washington Publication 609. (Reprinted in 1980)

Sitney, Lawrence, Paul Gray, David Phillips, and Eric Blinman
1995 Anasazi Pottery-Firing Technology: Los Alamos Firing Experiments, Phase I. Paper presented to Materials Science and Technology Group 4, Los Alamos National Laboratory.

Swink, Clint
1993 Limited Oxidation Firing of Organic Painted Pottery in Anasazi-style Trench Kilns. *Pottery Southwest* 20.

1996a Firing Anasazi Trench Kilns. In *Mesa Verde Waterline Replacement Project Phase III Archeological and Historical Studies*, by G.A. Ives, J.M. Brisbin, J.A. Bell, and C.L. Williams. Manuscript on file, Division of Research and Resource Management, Mesa Verde National Park, Colorado.

1996b Trash or Tools? Five Late Pueblo III Kiln Ceramic Assemblages.

Pottery Southwest 20. In press.

Swink, Tana
1996 The Effect of Freezing on Greenware and How It Could Have Affected Anasazi Pottery Production. *Pottery Southwest* 20. In press.

Toll, H. Wolcott, Thomas C. Windes, and Peter J. McKenna
1980 Late Ceramic Patterns in Chaco Canyon: The Pragmatics of Modeling Ceramic Exchange. In *Models and Methods in Regional Exchange*, edited by Robert E. Fry, pp. 95-117. SAA Papers 1.

Wilson, C. Dean, and Eric Blinman
1995 Changing Specialization of White Ware Manufacture in the Northern San Juan Region. In *Ceramic Production in the American Southwest*, edited by Barbara J. Mills and Patricia L. Crown, pp. 63-87. University of Arizona Press, Tucson.

THE VARIABILITY AND EVOLUTION OF PREHISPANIC KILNS ON THE PERUVIAN COAST

Izumi Shimada
Department of Anthropology
Southern Illinois University
Carbondale, IL 62901

ABSTRACT

The scarcity of archaeologically documented ceramic kilns has been commented upon often and has led to the widely held opinion that prehispanic ceramics in the Andes, like ethnographically documented cases, were fired predominantly in open pits. However, the past decade has seen a series of excavations at production sites with firing structures on the Peruvian coast, particularly on the north coast. Together they span some 2500 years of prehistory and show considerable variability in regard to construction materials, size, shape, capacity and overall design. This paper describes and examines the variability found among these firing structures within the broader contexts of production sites and technologies. The observed variability seems to be in part related to the vessel formation techniques employed, exterior finish desired, scale of production, and physical and social settings.

INTRODUCTION

With the abundance of elaborately decorated pottery in the ceramic stage of Andean prehistory, archaeologists have devoted the lion's share of their attention to stylistic and iconographic studies. While there has been important advances in stylistic dating and elucidation of rituals and associated ideologies, our research on the technical and technological dimensions of these ceramics has considerably lagged behind. Since the days of Sigvald Linné's classic *The Technique of South American Ceramics* (1925), such understanding has been based largely on ethnographic analogy and visual examination, combined with the study of molds in museum collections and small-scale replicative experiments (e.g., Carmichael 1990; Donnan 1965; Tello 1938). In his synthesis of South American ceramics, Willey (1949:142) was able to summarize what was known then about the whole subject of firing and associated features in a single, short paragraph. As a whole, earlier studies that touched upon the technological dimensions of prehispanic ceramics were "retrogressive", attempting to unravel the intricacies of the manufacturing processes from their products. This approach is inherently limited by the nature of the material evidence; i.e., artistically and technically superior examples without contextual data that had been looted from tombs and that were not representative of the vast majority of finished products.

This persistent, object-centered conception of craft goods has been a major barrier to understanding of manufacturing processes. It is not that production sites did not exist, but rather that there was no focused effort to locate them and conduct sustained, interdisciplinary studies. Only a few, definite production sites were known, fieldwork was geared toward other aims, personnel were often unfamiliar with key material clues, and there were no effective strategies for identifying production sites. However, the increasing thoroughness of field survey and surging interest in the social dimensions of technology of craft production over the past decade have resulted in a series of important discoveries and studies (e.g., Shimada 1994, in press). Today we can count at least nine prehispanic ceramic production sites on coastal Peru dating from the Formative (ca. 1000 B.C.) to the time of Inka Empire (ca. 15th to 16th centuries, A.D.).

It has been widely assumed that most ancient Andean ceramics were fired in the open or in pit kilns based on the widespread historic and modern firing practice and the rarity of documented prehispanic kilns. This paucity was interpreted as reflecting the pervasive use of simple firing structures without any permanent superstructures or enclosures. This circular argument has persisted in spite of archaeometric data and long-standing observations made on early ceramics such as those of the Cupisnique and Topará styles that suggest controlled firing in kilns (e.g., Menzel et al. 1964:212-213; see below).

Any investigation of firing technology must be conducted within a broader context of sustained, holistic study of ceramic production and use. I urge the adoption of an approach that integrates archaeological excavations of production and use contexts, multi-stage, replicative experiments of vessel formation and firing, and comprehensive archaeometric study both in and out of the field (Shimada 1994, in press; Shimada et al. 1994; Wagner et al. 1994a, b).

This paper is based on the results of recent and ongoing research centered at ceramic workshops with associated kilns on the coast of Peru (Fig. 1). The aim is to define variability in time and space of prehispanic ceramic firing structures on the Peruvian coast and offer some tentative explanations for the observed variability. The long period spanned by these sites allows us to elucidate some evolutionary trends. Relevant data are presented in chronological order from earliest to latest.

CUPISNIQUE KILNS (ca. 1000 - 700 B.C.)

For many years the Cupisnique ceramics on the north coast of Peru have won accolades for their representational, sculptural forms and subtle coloration achieved without use of slip paint. They represent the beginning of what Willey (1964) called North Peruvian Plastic Art Tradition. Larco (1941:35-36; 1948:17) argued that the subtle shades of gray seen in Cupisnique ceramics of the north coast of Peru were achieved through precise control of firing conditions using closed kilns. Additionally, recent Mössbauer spectroscopic analysis of Formative ceramics from the sites of Montegrande (upper Jequetepeque Valley, north

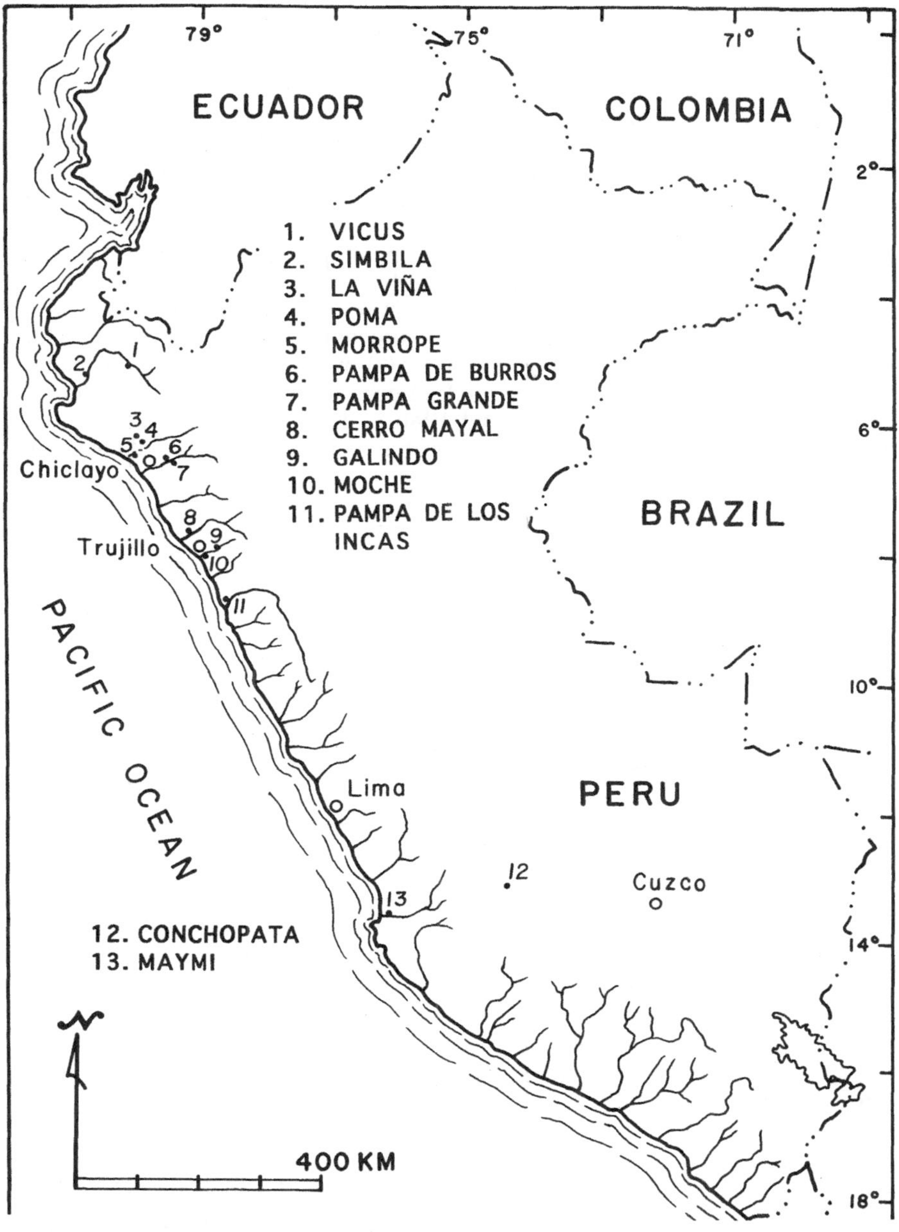

Fig. 1. Map showing prehispanic and modern ceramic production sites on the coast of Peru discussed in this paper.

coast), Chavín (North Highlands), and others showed that they were fired in semi-closed kilns that allowed for changing atmospheres (e.g., initial firing under reducing conditions followed by firing in an oxidizing atmosphere), relatively high temperatures (ca. 800° C up to 1000° C), and even heating (Salazar et al. 1986; Wagner et al. 1994a, b).

The use of just such semi-closed kilns for firing of Cupisnique ceramics was recently documented in the Poma Archaeological and Ecological Reserve in Batán Grande on the northern end of north coast of Peru (Location 4, Fig. 1). In 1985, examining the stratigraphy of the abandoned modern Poma Canal exposed by the severe 1983 El Niño flood, we found some 40 partial and nearly complete prehistoric kilns. The straight canal runs roughly northeast - southwest for about 4 km, measures 3-5 m in depth, and varies from 4 to as much as 50 m in width where there is severe erosion. The kilns are situated at depths between ca. 3.0 and 4.5 m below the surface (Fig. 2) in a 500-m long section of the canal (between 2.75 km to 3.25 km north of the La Leche River outlet). Half dozen of these were partially cleaned in 1985 and 1986 to determine their preservation, form and dimensions, and to collect samples of associated ceramics and charcoal.

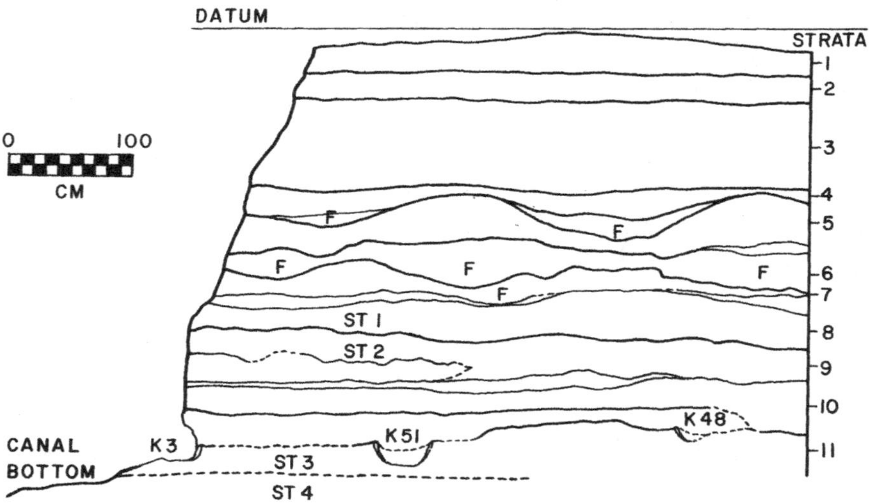

Fig. 2. Stratigraphy of the Poma Canal showing the superimposed positions of kilns. K refers to kilns, ST indicates Stratigraphic Positions, and F corresponds to cultivation furrows.

Additional survey and accompanying large scale excavations in 1989 revealed a total of 57 kilns within the same 500-m stretch (Shimada et al. 1990, 1994). Most were found within a 3-4 m wide area exposed along the base of both the east and west banks of the canal. While the considerable overburden hampered our ability to define the kiln distribution along the east-west axis, it is apparent that kilns tended to form clusters. For example, within a 10 x 6 m area cut into the west bank, our excavation revealed 10 kilns, some separated only 1-2 m (Fig. 3). We believe there are literally dozens more to be discovered in this area.

The kilns have been found at three distinct Stratigraphic Positions, 2 through 4 (counting from the top; Fig. 2). There may be additional kilns in Stratigraphic Position 1. Thick silty layers overlying Stratigraphic Positions 2 and 3 are dark and full of reddened kiln fragments and charcoal bits. These layers span much of the aforementioned 500-m stretch of the canal, underscoring the presence of numerous kilns and intensity of firing activities. The kilns in Position 3 are most numerous, largely due to the erosive effects of canal water and the some 20 excavated kilns in this position are are most intensely studied thus far. Dating of charcoal samples recovered from the kiln bottoms and associated Chólope style sherds suggest that kilns at Positions 2, 3 and 4 date to ca. 470-850B.C., 620-840 B.C. and 960-1260 B.C., respectively (95% confidence level; Shimada et al. 1994:82-83).

Each Stratigraphic Position is an extensive, clay-rich slack flood deposit (Craig and Shimada 1986). Petrographic thin-section and x-ray diffraction analyses show that local clays contain sufficient kinds, sizes, and quantities of mineral inclusions (chlorite, kaolinite, mixed-layer minerals, as well as biotite, feldspars and amphiboles) to make them “self-tempered” (Killick n.d.; Wagner et al. 1994a,b). Further, petrographic thin-section and neutron activation analyses of samples of kiln lining and associated ceramics amply demonstrate that they were made from local clays (Shimada et al. 1994; Wagner et al. 1994a,b). In other words, ancient potters simply took advantage of extensive, readily exploitable beds of good clay. The kilns were built simply by digging pits into each clay bed.

The 57 kilns documented thus far were quite uniform in overall design, construction material, and technique, while exhibiting some degree of variation in size and shape (Fig. 4). The classification below is based on functional attributes:

I. Single-Chamber Kilns - one definite and two possible examples are known, though it is possible that they are the poorly preserved principal chambers of the double-chambered kilns described below.

II. Differentiated Double-Chamber Kilns - differentiated in the size, shape and depth of their component chambers:

A.Single Chimney

1."Keyhole-shaped" (Fig. 5) - spherical principal chamber with elongated antechamber (the length of the antechamber is greater than that of the principal chamber).

2. "Pear-shaped" (Fig. 6) - large ovoid principal chamber and considerably smaller antechamber (the length of the antechamber is less than that of the principal chamber). These are the most common and widely distributed.

B. Multiple Chimney (Fig. 4) - a variant of the "Pear-shaped" kiln with three chimneys. Only one case is known.

III. "Figure-8" Kilns - two large chambers of equal size and shape with relatively flat bottoms; two cases are known.

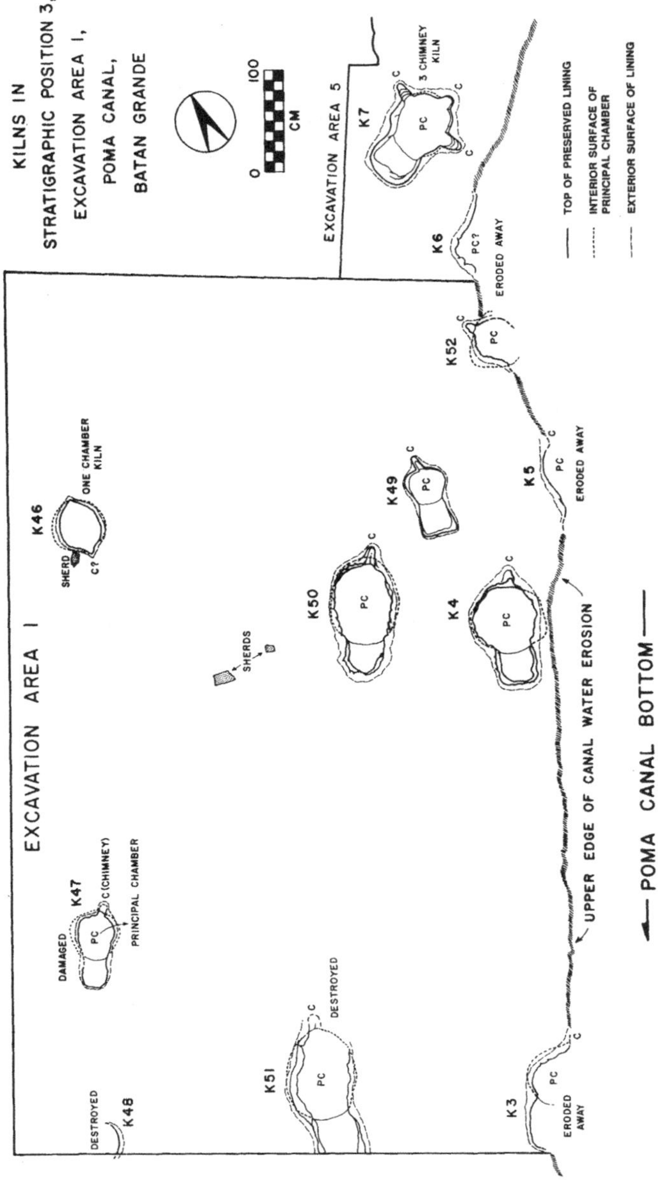

Fig. 3. Clustering of kilns documented in Excavation Area 1, Poma Canal.

Like any classification, not all kilns are readily assignable to the above categories and I make no claim that the above reflects the indigenous kiln classification. Rather, this scheme helps us understand the possible functional significance of observed variation in the size and shape of the kilns. The following are the basic features observed among the excavated "double-chambered kilns":

(1) These kilns consist of two overlapping pits, a larger, deeper principal chamber and a smaller, shallower antechamber, dug into an in situ clay-rich flood deposit. Kilns show varying degrees of constriction at the waist. The ranges of key measurements of these kilns are presented below in Table 1:

Table I: Kiln Measurements

	Principal Chamber	Antechamber
Width, range	40-70 cm	25-40 cm
Typical width	60 cm	30 cm
Depth below ground level	70 cm	50-60 cm
Typical depth	35-45 cm	
Interior length, range	760-140 cm	
Typical interior length, range	100-120 cm	
Chamber capacity	0.15-0.50 m^3	
Typical capacity, Keyhole kilns	0.20-0.25 m^3	
Typical capacity, Pear-shape kilns	0.40-0.45 m^3	
Chimney opening, range	4-15 cm (at neck)	
Chimney opening, typical	12-13 cm	

(2) Nearly all principal chambers show either a vestige of or a well-preserved vertical, flaring chimney that projected slightly above the ground at its distal (north) end. One large kiln had two additional chimneys at the widest points of the principal chamber (Fig. 4). Chimney openings vary from ca. 4 to 15 cm (at its neck), but more commonly ca. 12-13 cm wide. Some chimneys start close to the bottom of the principal chamber, while others are built at the point where the chamber wall begins incurving.

(3) Above ground level, kilns have converging side walls as extensions of the curvature of walls of the two overlapping pits dug into the clay bed. For the principal chamber, the walls are estimated to have reached ca. 12-15 cm above the ground level based on the inward inclination of the walls (ca. 30-40°) and the amount of fallen lining. The converging walls left a longitudinal opening 20-30 cm wide at the apex (Fig. 5). In other words, the walls created a dome-like structure and a semi-closed environment. The side walls of the antechamber were more vertical (curving in by 10-20°) and limited in height (perhaps ca. 10 cm). The superstructures, their lower extensions, and the interior linings of the two chambers are constructed of a mixture of local clay and locally abundant *algarrobo* leaves (*Prosopis pallida*). Preserved parallel finger grooves indicate that the lining was firmly applied by hand. The resultant vertical corrugation

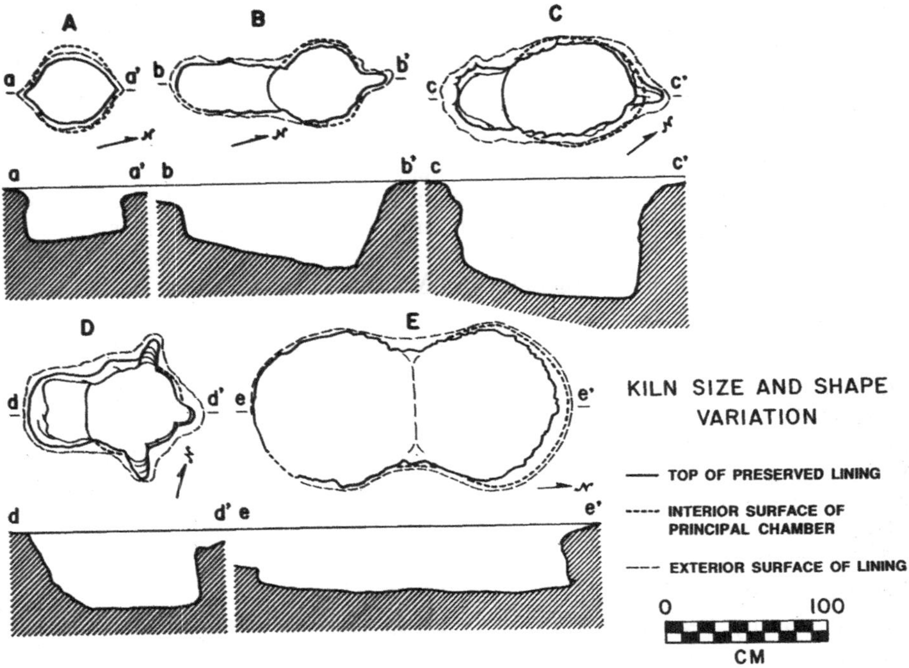

Fig. 4. Classification of the excavated kilns based on variation in size and shape.

probably strengthened the lining against the weight of the dome and prevented damage from repeated expansion and contraction. The lower halves and floors of the chambers are smoothed but do not have linings. The thickness of the lining is usually the greatest at ground level to support the weight of the "dome" and at the neck of the chimney, measuring up to 15 cm. The inferred dome fragments found within the principal chamber are considerably thinner, typically about 2-3 cm, while those of the antechamber are even thinner, often only 1-2 cm. With the possible exception of one or two chambers, there is no evidence of relining.

(4) Kilns were oriented north-south to northwest-southeast (with the principal chamber on the northern end), matching the prevalent diurnal wind directions of the region.

(5) Bases of *quincha* or wattle-daub walls and a few postholes were found just a few meters north of Kiln Cluster 3. The architectural configuration indicates that kilns were situated outdoors. The presence of hematite with use-wear and polishing stones in Excavation Area 1 suggests that vessels may have been made within several meters of the kilns.

(6) Each kiln is associated with only a limited number of sherds. Most commonly, a few to a handful of relatively large, thick body sherds of big

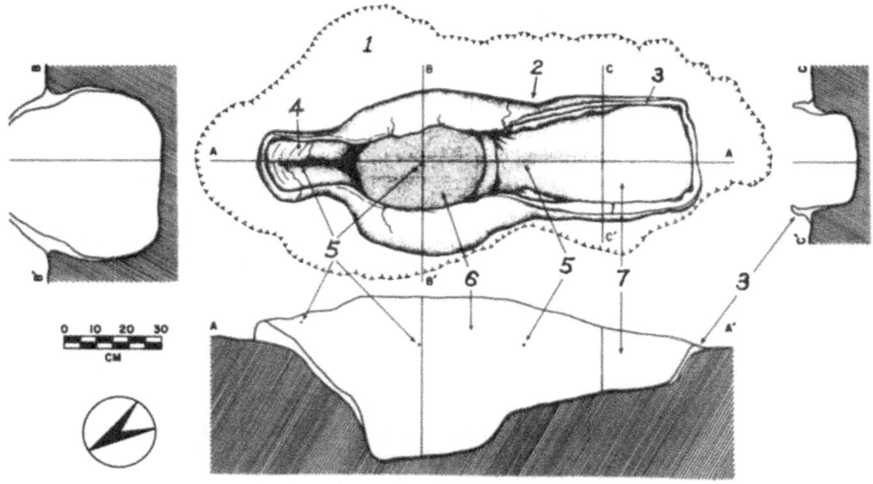

Fig. 5. Various views of the full-scale, replica keyhole-shaped kiln built for our refiring experiments: (1) heat-reddened area, (2) constricted waist, (3) lining, (4) chimney, (5) thermocouples, (6) main chamber, and (7) antechamber.

Fig. 6. 2,800 year-old Kiln 38 (Pear-shaped) used for refiring experiments.

utilitarian storage urns or jars were found inside the principal chamber mixed with fallen fragments of side walls. Fragments of open bowls and short-neck jars, some showing spalling, were also found inside the kilns as well as on associated floors (corresponding to the top of clay beds).

(7) Kilns have either a reddish or grayish color, with the most notable discoloration seen in the upper reaches of the chimney and walls of the principal chamber. The necks of some chimneys are bright pinkish-red in color and hardened like a ceramic vessel. The waist constriction also shows some discoloration. In contrast, the lower portions of the principal chamber, including the floor often show gray coloration. The antechamber as a whole is noticeably less discolored and hardened by heat than the principal chamber. In fact, the mouths of some kilns show no discoloration at all, making it difficult to define form. In addition, the principal chamber almost invariably contains stratified, compacted ash and charcoal bits at the bottom, while the antechamber is largely clean. This differential deposition may be partly due to the fact that the floors of many antechambers slope down toward the principal chamber. Charcoal lumps were mostly less than 1 cm in width and length, though a few were nearly 10 cm in diameter. Microscopic examination of charcoal bits showed the tight irregular grain typical of *algarrobo,* an indigenous tree with hard, dense, resin-rich wood, the popular choice as domestic fuel and charcoal-making prehistorically and historically.

The preceding informs us much about the working and use-history of the kiln. The observed pattern of discoloration and hardening agrees well with our expectations. For example, the design of these kilns suggests that fuel was ignited at the mouth of the principal chamber (i.e., constricted waist) or in the antechamber and pushed into the principal chamber where much of the fuel would be. Thus, the antechamber would not have been subjected to intense heat. Once the fuel in the principal chamber was lit, the spherical shape and "lid" atop the dome of the pricinpal chamber would have been effective in both horizontally and vertically disseminating the flame and heat to evenly fire the vessels inside. At the same time, the open chimney would have created a strong updraft within the kiln, effectively heating and discoloring the chimney mouth. As detailed below, these expectations were confirmed independently in our replicative firing experiments.

The surface color of the kiln lining is believed to reflect prevailing atmospheric conditions at the end of the last firing. Thus, the gray discoloration that characterizes the 4 constituent kilns of Kiln Cluster 3 indicates that a reducing atmosphere prevailed. The fact that the linings of these kilns are gray in color throughout their thickness suggests that they were perhaps reserved for obtaining a gray to black finish on the pottery. On the other hand, the 10 kilns composing nearby Kiln Cluster 1 are all reddish in color, suggesting their use for firing in an oxidized atmosphere. Other notable differences exist between kilns of Clusters 1 and 2: the former are generally large (including thick lining) and have relatively small, low antechambers (i.e., “Pear-shaped” kilns), while the latter are considerably smaller, have a rather accentuated "Keyhole" shape with relatively long, elevated antechambers. In other words, the former appear to have been used in relatively large-volume firing in an oxidized atmosphere, whereas the latter were employed in small-volume reduction firing.

Various lines of evidence presented above also suggest that the desired firing conditions could be readily achieved by changing air flow through the chamber mouth and top as well as the chimney. It is argued that the large, thick sherds found buried within the principal chambers are parts of a removable lid for the chimney and/or dome-top opening. The opening would have allowed easy access to the chamber for loading vessels and fuel, and, following firing, removal of fired vessels and cleaning. Once the opening was closed with sherds, the resultant dome would serve as a refractory barrier and establish a convection cycle essential for even firing of vessels within the chamber. The chimney, then, becomes crucial for proper operation of this semi-closed, updraft kiln. Its special importance is apparent in the care with which it was built and its intense heat discoloration. Though these ancient kilns were well made, the dome-like superstructure of the principal chamber was vulnerable. Once collapsed, it would have been difficult to repair and probably led to the abandonment of the damaged kiln.

Inferences regarding the specific design features and workings of these Formative kilns were tested through a step-wise series of replicative firing experiments carried out in and out of the field in 1989 and 1990 (see Shimada et al. 1994; Wagner et al. 1994a,b, in press for details). The experiments began with a 2/3 scale model and proceeded to a full-size replica kiln built at the bottom of the Poma Canal next to Kiln Cluster 3 (i.e., Stratigraphic Position 3; Fig. 5). It faithfully reflected our understanding of the structural design and construction materials and techniques of the Formative kilns. Local raw materials were used to the maximum extent possible, including use of clay from Stratigraphic Position 3 to line the kiln. Fuel consumption and internal temperatures were monitored. Experiments were refined over time and culminated with refiring of a 2800-year old "Pear-shaped" kiln in Stratigraphic Position 3. For this last, temperature variations at five different spots within the principal chamber of the kiln was recorded with five Ni/CrNi thermocouples and digital thermometers. In addition, clay briquettes were placed at roughly the same spots. Ongoing laboratory analysis of the briquettes and samples of the lining of this kiln as well as replica ceramics fired during the experiments is being carried out by Ursel Wagner, a German applied physicist, and her colleagues at the Munich Technical University. The analysis centers around Mössbauer spectroscopy, which examines the behavior of iron compounds in clay to elucidate firing temperature and condition. A complementary study of the heat transfer process of the clay used in the kilns and ceramic manufacture is also conducted (Wagner et al. n.d.). This battery of archaeometric analyses is providing independent confirmation of insights and observations of the actors and observers involved in these field refiring experiments.

These experiments clearly show that both "Keyhole" and "Pear-shaped" kilns function in a highly predictable manner and are capable of attaining a firing temperature of ca. 800° C. Mössbauer spectroscopic analysis indicates that most of the properly fired ceramics from the Poma Canal and nearby, contemporaneous sites had been fired at about 800 °C (Wagner et al. 1994a, b).

Figure 7 for example, shows the temperature readings recorded during a refiring of the Formative kiln in an oxidizing atmosphere with 10 replica vessels and 13 kg and 5 kg of *algarrobo* and *zapote* (*Capparis ovalifolia*) branches, respectively, as fuel. The load is far below the capacity of the kiln which is ca. 0.5 cubic meters. Nonetheless, the figure shows quite a rapid and even rise in temperatures to ca. 700-750° C at various spots within the firing chamber. The maximum temperature of ca. 800° C was recorded near the bottom of the chamber some two hours into the experiment when hardwood fuel had turned into red-hot charcoal. Following that, temperature declined gradually and quite evenly at most locations within the chamber. We suspect a higher temperature could be achieved by loading with more hardwood fuel.

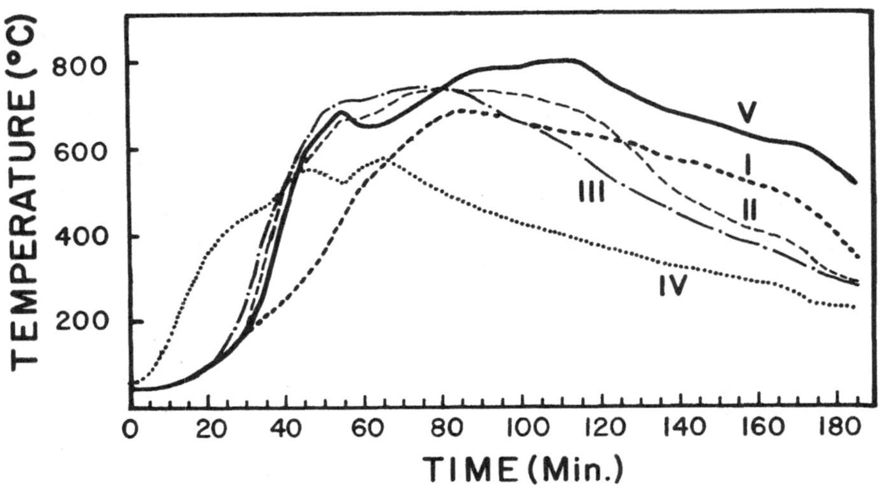

Fig. 7. Changing temperatures recorded by five thermocouples placed within Kiln 38 plotted against elapsed time.

In the above experiment, the dome-top opening was closed with large sherds and mud after loading. The chimney and mouth of the firing chamber were left open throughout the firing. Other experiments amply showed how readily reducing firing conditions may be attained by partially or fully closing the chimney with one or two sherds. The chamber mouth can also be closed in a similar manner. In one experiment, the waxy green leaves of *bichayo* (*Capparis cortata*), an abundant, indigenous shrub, were added to yield the shiny, black coating sometimes found in Cupisnique bottles. In this case, the leaves were loaded into the firing chamber through the mouth toward the end of firing. Overall, our experiments illustrate that the Cupisnique kilns behave predictably and are quite manageable. Visually, the quality of the finished wares, including tone and hue of surface color and color and hardness of paste, was comparable to that found in the vessels excavated at Poma Canal and various local Formative sites. Both the replica and original Cupisnique vessels were made with a heavily sand-tempered clay mixture that could withstand the effects of the rapid temperature rise during the early stage of firing.

MOCHICA AND NASCA PIT KILNS (ca. A.D. 450-750)

Four Mochica ceramic production sites on the north coast have been excavated in recent years (Fig. 1): (1) Cerro Mayal, a natural promontory near the major Mochica population center of Cerro Mocollope in the mid-Chicama Valley; (2) Moche, the early capital of the Southern Mochica in the lower Moche Valley; (3) Pampa Grande, and (4) Galindo. The last two represent the late capitals of the Northern and Southern Mochica polities located at the necks of the Lambayeque and Moche valleys, respectively. Associated radiocarbon dates and ceramics suggest that ceramic production at the first two sites spanned ca. A.D. 450 to 750 (Mochica Phases IV-V), while the last two were restricted to ca. A.D. 550 to 750 (Phase V). In addition, Pampa de los Incas, the Mochica regional center in the lower Santa Valley (ca. A.D. 450-550; Phase IV) has yielded molds and other remains suggesting on-site ceramic production (Wilson 1988). It comes as no surprise that there would be at least one ceramic production locus within or near major population centers.

A large quantity and variety of production debris are found at the east end of Cerro Mayal where Glenn Russell and his colleagues (1994) have been working since 1989. The ceramic production area and associated waster piles together cover an area ca. 185 by 90 meters along a small gully. Much of this area has archaeological deposits mixed with ash and charcoal bits to a depth of 1.0 to 1.5 meters below surface. On the basis of their survey and test excavations, they defined spatial differentiation of inferred distinct but related activities. The lower reaches of the gully yielded a high density of varied production debris including mold fragments, unfired vessels and figurines, some of which were already slipped and even polished and painted, and lumps of shaped and cut prepared clay mixture. Vessel formation and decoration are indicated in the lower gully.

Such debris was considerably less dense upslope even taking into account the expected downslope movement. In this upper gully area, Russell et al. (1994:211-212) found the remains of firing structures, including an inferred pit kiln. This oblong, shallow depression, ca. 2.5 m in length and 25 cm in depth, was not placed within any well-defined or protected structure but rather was dug into accumulated (presumably outdoor) midden. The pit showed intense heat discoloration. Charcoal bits found at the bottom of the pit appear to be of *algarroobo*, which is believed to have been the primary fuel. The kiln's location on the sloping side of a gully through which the prevailing winds blow would have provided a good draft during firing. That the kiln achieved high temperatures is indicated by the density of "over-fired, warped, and vitrified sherds as high as 62 /m^2" (Russell et al. 1994:209). Salazar et al. (1993) recorded temperatures of over 1050° C (using Ni/CrNi thermocouples) during firing of a similarly sized and shaped modern pit kiln in Mórrope (some 170 km to the northeast of Cerro Mayal) that also utilized *algarrobo* hardwood as the main fuel. These density figures are important because they imply excessive temperature and, more broadly, poor control of firing conditions.

A layer of white ash several centimeters thick was found at the bottom of the depression. Microscopic examination revealed the ash consisted of the burnt stems of locally abundant *cola de caballo* (horsetail; *Equisetum giganteum*; Russell et al. 1994:211), which is a primitive plant with segmented, hollow stems and cone-like spikes and grows up to one meter high in sandy, alluvial soil or along stream banks and other moist soil. Stems have abundant, embedded silica. Several small fragments of burnt clay with impressions of horsetail stems were also found in the kiln. Though no definite remains of kiln superstructure were recovered in the Cerro Mayal pit kiln, Russell et al (1994:212) suggest that the clay lumps with horsetail stalk impressions represent broken pieces of an ephemeral kiln cover. In addition, they (1994:212) believe that the absence of a distinct layer of carbon at the kiln bottom reflects the placement of hardwood fuel around and between vessels being fired. Thus, they (1994:212) visualize firing preparation as follows: "... it appears that the kiln consisted of a pit in which a combination of fuel and ceramic objects was placed using a mixed firing approach. Finally, the kiln was covered, probably with a combination of sherds, [horsetail], and possibly sealed with mud prior to firing."

Horsetail remains were also found in two circular pit kilns excavated by Martha Anders and her colleagues (in press) at Maymi, a contemporaneous late Nasca ceramic workshop in the lower Pisco Valley over 750 km to the south of Cerro Mayal. V. Chang (personal comm. 1996), a Peruvian potter long interested in prehispanic pottery technology, observes that traditional potters on the south coast of Peru place a layer of crushed, dry cow dung mixed with horsetail atop and below vessels to be fired in their pit kilns. it appears that sillica in the plants serves to *aerate* the dung so that it burns evenly and thoroughly. This is a quite plausible explanation. At Maymi and Cerro Mayal, llama dung could have been readily gathered because of the animal's abundance and common voidance behavior (Shimada and Shimada 1985).

Data from Galindo complement the preceding discussion. A shallow (depth not indicated), circular pit kiln some 7 m in diameter was found in the northern margin of the extensive urban site (ca. 6 km^2) in association with a small ceramic workshop (15 m^2 in area; Bawden 1977:187-198). The kiln was filled with ash containing wood, llama dung, and large fragments of coarse, utilitarian vessels. In and around the pit were 65 broken molds of neck jars, double-chamber bottles and figurines. Also near the kiln was " "storage pit", 1.5 m in diameter and 1 m in depth, containing cane and wood mixed with llama dung, apparently fuel intended for the kiln. Bawden (1977:193) suggests that the vessels were first packed with the above fuel and then, after ignition, covered with a layer of earth. Given that the mold fragments are of relatively small vessels and figurines, the large sherds were likely to have been used as parts of a refractory kiln cover or pedestals to elevate vessels off the kiln floor.

Reconstruction of kilns and the placement and nature of their loads, however, is not a simple matter. Russell et al. (1994:220) believe that Cerro Mayal produced primarily small decorated vessels (e.g., flaring vases, neck jars, stirrup-spout bottles, closed bowls), figurines, and musical instruments (trumpets and rattles) by combining hand- and mold-made components. Utilitarian vessels (large storage ulrns and cooking *ollas*, vessels with wide mouths, globular bodies with or without short necks) were apparently produced elsewhere. Some forms made at Cerro Mayal were not restricted to either elite or ritual contexts. However, it is not known whether widespread neck jars were produced and fired together with more restricted bottles. Likewise, whether and how product and kiln variability are related have not yet been defined. At a more specific level, many sherds found mixed or lying atop the ash layer in the pit kiln excavated at Cerro Mayal were overfired and brittle. It is difficult to define whether the sherds were part of the kiln cover or pedestals for vessels being fired.

The excavations at Cerro Mayal and Moche yielded possible evidence of another kind of Mochica kiln, one with low adobe brick walls and a roughly rectnagular shape. The discovery of numerous intensely reddened adobe bricks recycled in constructions near the aforementioned pit kiln at Cerro Mayal led Russell (personal comm. 1996) to infer that these adobe bricks may have been used originally to partially enclose kilns. In fact, a row of intensely discolored adobe bricks set in clay mortar (ca. 1.5 m long, 20 cm. wide and 25 cm. high) associated with a similarly discolored surface underneath was found in the same area.

At Moche, some 150 m south-southwest of the high status cemetery at the west base of the major, multi-level adobe brick temple of Huaca de la Luna, a damaged late Mochica (primarily Phase IV) pottery workshop was excavated (ca. 240 m^2 area) by a team from the National University of Trujillo in 1993 and 1994 (Uceda and Armas in press). This general location had been suspected of being a major craft production area since the pioneering work of Max Uhle (1913, n.d.) in 1899-1900. T. Topic (1977:156, 406-410; 1982:275-276) found considerable quantities of wasters, broken figurines and figurine molds in the same area. The University of Trujillo team defined three occupational phases in the excavated workshop, with all floors and associated fills yielding a wide variety of ceramic manufacturing implements, features, and debris. For example, over 1,000 whole and fragmentary mold matrices and molds were recovered from the excavation. Uceda and Armas (in press) conclude that the production of this workshop was devoted to high quality products for elite consumption and ritual purposes even though almost all known Mochica style pottery forms were manufactured there. Further, they argue that activities within the workshop encompassed all steps of production, including preparation of clay mixture, mold matrices, and molds, vessel formation, firing and storage of finished products.

In respect to kilns, they exposed part of a circular pit kiln belong to the first phase workshop. It was in a second phase workshop that a possible rectangular kiln were exposed. Much of the north end was occupied by two parallel walls (ca. 1 m apart) of rectangular adobe bricks set in mortar that had been thoroughly and intensely reddened and were associated with soil mixed with ash. The better preserved south wall measured ca. 1.5 m long, 25 cm. wide, and 35 cm high -- much like the discolored wall at Cerro Mayal described above. These walls are believed to have enclosed partially a rectangular, above-ground ceramic kiln. At the same time, there is no evidence of a superstructre to indicate that the inferred kiln was fully enclosed. Large, removable sherds may have been used to seal the kiln. The east and west ends of the inferred kilns are not well defined, but may have been left open for setting of the fire and removal of ash.

The above cases of possible rectangular kilns raise more questions than answers. For example, why are they rectangular as opposed to circular or oval? What advantages are offered by this form and low walls?

One factor to consider in regard to these questions is the physical and social settings of the inferred kilns. Though the Moche excavation was limited to a portion of one workshop, the surface distribution of mold fragments and architectural patterns of nearby excavated areas suggest there are still other workshops to be excavated. In other words, we are speaking of craft production within an urban setting where space and control over movement may have been at a premium. Could the shape of the inferred kiln and its placement have been conditioned largely by its setting? The rectangular shape could be better accomodated in the restricted area at the north end of the enclosed workshop. Its walls would keep the fuel and large sherd cover in place and facilitate cleaning of the pit after firing by keeping ash from spreading. The walls would also allow different sized vessels to be stacked within the kiln. Such loading may facilitate maintaining an oxidizing atmosphere. In other words, for the surface area it occupies, this kind of kiln has a relatively large loading capacity and, at the same time, reduces the environmental impacts of firing.

Some traditional Muchik potters of Mórrope (south edge of the Sechura Desert) and Simbilá (lower Piura Valley on the far north coast) erect low adobe walls (up to 90 cm high but more commonly 30-35 cm high or 3 to 4 courses with little or no use of clay mortar) around the perimeters of their circular or oval pit kilns leaving one or two narrow openings for setting the fire (Fig. 8; see Figs. 1, 2 in Salazar et al. 1993:689, 690). The walls are said to serve the purpose of stabilizing the stacked vessels, fuel and sherd cover, and facilitating post-firing cleaning. In general, these modern potters have few spatial limitations in regard to the size, shape and relative placment of their kilns.

It remains uncertain to what extent the above discussion regarding the inferred rectangular kiln at Moche is applicable to kilns at Cerro Mayal. For example, why do circular pit kilns and rectangular kilns with basal walls coexist at Cerro Mayal? At both sites the excavations conducted thus far have been limited in extent and archaeometric studies of the kilns and asociated ceramics for reconstruction of late Mochica firing technology have just began. Given the quantity and variety of production debris, both sites have potential for a full reconstruction of late Mochica ceramic manufacturing technology and organization of production. It should be recalled, however, that the ceramic production area at Cerro Mayal is physically detached from the major urban settlement of Cerro Mocollope, while the inferred aggregation of ceramic workshops at Moche appears to be set within the extensive urban layout of the site, close to the seat of inferred religous leaders at Huaca de la Luna (Uceda and Armas in press). Thus, we would expect significant differences in the organization of production.

Workshops at Moche may well have been characterized by the "additive" or "modular" approach found in various aspects of the Mochica culture. For

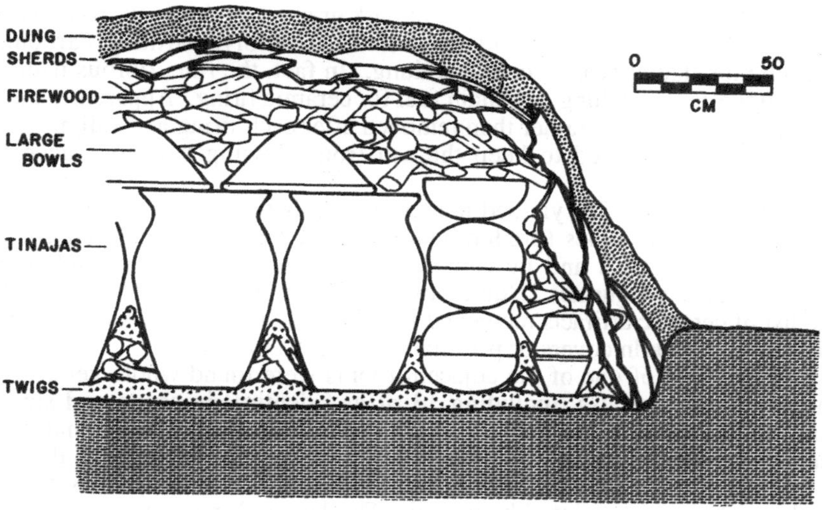

Fig. 8. Cross-sectional view of a typical modern pit kiln in Morrope.

example, Mochica monumental adobe structures were essentially constructed using tall, standardized, vertical columns of adobe bricks mortared internally but not joined with adjacent columns (e.g. Kroeber 1930; Moseley 1975). In craft production, this approach meant a set of supervised, similarly organized and equipped workshops in the same general area (contiguous or dispersed but readily accessible to each other by means of shared corridors or patios) engaged in the same craft, producing the same whole objects or complemetnary parts to be assembled elsewhere (Shimada 1978, 1994). Procurement of basic ingredients and distribution of products are believed to have been supervised. We are speaking of production by the masses that allows some degree of autonomy at the level of individual production units. The approach may well have fostered what Stone-Miller (1995:88) calls "personalisms" of Mochica ceramic art.

Data from the late Mochica capital of Pampa Grande (6 km^2 in extent with an estimated population of 10,000) offers another perspective on ceramic kilns and production in an urban setting. The evidence for ceramic production comes from a cluster of small, contiguous rooms that together occupied an area of ca. 10 x 15 m toward the western edge of a sector near the center of the site. This area housed craft workshops, a formal storage facility, food and beverage preparation areas and administrative structures, all interlinked via corridors and streets (Shimada 1994:195-200). The activities and products of this sector are believed to have served the needs of elite residents in adjacent sectors.

A small, rectangular masonry room (ca. 2 x 3 m) and two adjacent rooms are of particular interest. The masonry room is believed to have been a kiln, though evidence is not conclusive. Many of the exposed wall stones are blackened and/or brittle, suggesting exposure to fire. The room contained a nearly meter thick, homogeneous layer of dark gray ash and soil mixture. Within this layer were tiny charcoal bits and small, crumbly, partially carbonized organic lumps that appear to be llama dung. In fact, the fine, fibrous nature of the ash suggests llama dung as a main fuel for ceramic firing. Apparently the ash was allowed to build up, rather than removed after each firing and soil may have been intentionally mixed to keep the ash from blowing between successive firings.

The ash deposit also yielded numerous sherds representing a wide variety of vessel forms and types (predominantly reduced ware serving and cooking vessels, such as plates and bowls), as well as stone burnishers and broken stamps, molds, and figurines. Numerous, large urn fragments that seem intrusive to the list of ceramic artifacts recovered from the ash deposit may well have been used as refractory kiln covers or pedestals for objects to be fired.

On the floor of one of the adjacent rooms were found whole vessels with no apparent signs of use, including 13 *ollas* identical in shape and size but distinct in decoration. Other vessels (e.g., bowls) matching the form and size of vessels represented in sherds from the inferred kiln also were found on the floor. The adjacent room on the other side yielded whole and broken potters' plates, a stone burnisher and a pointed bone tool. The last tool may have been used in decorating the bowls and *ollas* in the other room as the pointed end fitted perfectly in the grooves and punctated designs.

Overall, these lines of evidence argue that these three contiguous rooms were used at least for the last stages of ceramic manufacture, including decoration, reduction firing and storage of finished vessels. Nearby unexcavated rooms may have served for clay storage and preparation.

There is at least one documented historical case of a small rectangular kiln (estimated 3.5 x 2.5 m) enclosed on all four sides by adobe walls. In 1907, Heinrich Brüning photographed a kiln in Monsefu (lower Lambayeque Valley; his Photo No. 739; Schaedel 1988:116) with unplastered enclosure walls ca. 1.2 m high. Little mortar was used between courses and vertical wooden posts were erected to shore up the walls. Numerous large sherds covered the full load ready to be fired. It is described as having been used to fire *ollas*. Ceramics are no longer made in Monsefu today and such kilns are no longer found on the north coast.

Rectangular kilns are not restricted to the north coast. Two contiguous, rectangular kilns with vertical masonry walls (ca. 2.5 x 6 and 3 x 5 m in dimensions) filled with ash, a thick layer of large sherds and burnt soil are also reported from the contemporaneous Wari site of Conchopata near the city of Ayacucho in south-central highlands (Pozzi-Escot et al. 1994:289). As in the case of Cerro Mayal, a circular pit kiln (ca. 3 m in diameter and 60 to 80 cm in depth) that had been heat-reddened and filled with ash is also reported from Conchopata (Pozzi-Escot et al. 1994:289). At both sites, however, functional and chronological relationships between the two types of kilns remain to be clarified.

Departing the north coast, we find no published report of the prehispanic kilns on the coast, with the notable exception of the late Nasca site of Maymi. Evidence of ceramic production comes from the central portion of the elongated site (river terrace) that covers an area of ca. 650 x 180 m (Anders et al. 1994, in press). Due to the aridity of the region, preservation of organic substances is excellent. Thus, recovered production remains include not only numerous sherd scrapers, stone burnishers, unfired clay, and ceramic turntables, but also paint brushes made with human hair as well as worn corncobs and cotton rags used for surface smoothing. Surface-collected and excavated artifacts suggest spatial differentiation of successive phases of ceramic production starting with clay storage and preparation, vessel formation, vessel decoration and finishing to storage of finished vessels. The first area is closest to the inferred clay source exposed along the south side of the site.

The differentiated areas as a whole formed a semi-circle (over 40 m across) around two excavated kilns only 3 m apart and measuring ca. 1 and 3 m in diameter and 30 and 40 cm in depth, respectively. The walls of both kilns show intense reddening to a considerable depth, suggesting numerous firings in an oxidizing atmosphere. Large plainware body sherds were found still lining the walls and edges of the kiln floors. Compacted ash adhered to the walls and the partially charred remains of *cola de caballo* were found on the floors. No evidence of a kiln superstructure was found. Overlying these remains on the floors, Anders (1990) found clustered fragments of fine decorated vessels of varied form that suggest intentional breakage as offerings at the time of or following cessation of the kiln use. It is likely that there are additional kilns near the two known ones, given the extent and quantity of production debris and the inferred workshop industry level of production.

PROVINCIAL CHIMU AND INKA KILNS (ca. A.D. 1375-1532)

The 1991 surface survey in Pampa de los Burros along the ancient Taymi Canal near the neck of the Lambayeque Valley revealed a ceramic workshop dating to the period of Chimú domination of the valley around A.D. 1375 to 1470 (Tschauner et al. 1994). The workshop consisted of a large circular pit kiln and adjacent complex of 18 small, agglutinated masonry rooms (ca. 20 x 40 m). It occupied the northeast margin of, but was clearly detached from, a small settlement of residential character, LV-S166 (ca. 80 x 260 m).

The associated kiln is situated within a rectangular area partially protected on three sides by low masonry walls (Tschauner et al. 1994:370-371). The reddened soil forms an irregular shape approximating a circle ca. 9 m in diameter. Its 5-10 cm depth is filled with dark ash, numerous bits of *algarrobo* charcoal, and nearly 30,000 small sherds, predominantly reduced wares, representing a wide range of Chimú ceramic forms (total weight of 207 kg). Within this large kiln is another intensely reddened pit, 6.5 by 4.5 m and 40 cm in depth, partially filled with burnt and compacted animal dung that the excavators (Tschauner et al. 1994:372) consider to be of guinea pig (*Cavia porcellus*). At least one side of the smaller pit appears to have been lined with stones and large sherds.

Much of what were recovered in the 1992 excavation of the workshop came from in and around the large kiln, including, among other artifacts, 35,000 sherds, stone burnishers, unfired clay lumps, ceramic paddle fragments, whole and broken molds, and mold matrices. These remains show that the vessels produced in this workshop were primarily reduced ware *ollas*, stirrup-spout bottles, plates and double-chambered jars. Tschauner et al. (1994:373) observe appropriately that the workshop produced not only these mold-made vessels but also pottery making tools such as paddles, molds and their matrices.

Tschauner et al. (1994:374) believe another smaller pit kiln (only partially exposed; ca. 2 m on one side) was placed near a corner of a roughly square masonry room (ca. 7 m to a side) immediately south of the large pit kiln. The inferred kiln showed heat discoloration, associated ash deposit, charcoal bits, and carbonized dung.

Evidence from surface collection and excavation showed that only two rooms at one end of the 18 interlinked room complex were definitely used for ceramic production. Some of the remaining rooms appear to have been used for domestic activities. Overall, data from the Pampa de los Burros workshop point to domestic industry production of relatively small reduced ware vessels using one or two simple pit kilns.

The use of pit kilns is also apparent at the extensive provincial Inka ceramic workshop at La Viña in the lower Leche Valley on the north coast (Hayashida 1994, personal comm. 1995). Hayashida's excavation revealed a series of shallow, heat-reddened, circular depressions filled with ash and charcoal bits and associated with mold fragments, misfired sherds, unfired vessels and clay lumps. We await publication of her report.

DISCUSSION AND CONCLUSIONS

This survey of recent investigations into prehispanic ceramic production on the Peruvian coast illustates that today there are a handful of long-term studies focused on ceramic workshops of different periods and cultures. These studies reveal much more variability in size, shape, design, inferred operation and social and physical settings of kilns than previously assumed. We have seen semi-closed, double-chambered kilns with chimneys, simple circular pit kilns, rectangular kilns with masonry walls on all four sides, as well as inferred rectangular kilns with basal adobe walls on two sides.

The Cupisnique semi-enclosed kiln offered a considerable degree of control over firing conditions. It could be used as an open or closed system even within a single firing episode. Why, then, do we not see these kilns more widely used? What happened to these kilns?

A tentative explanation may be offered from the vantage of a long-term perspective. There was dynamic interplay between kilns and firing technique, on the one hand, and the level of production demand and the utilitarian and symbolic significance of ceramics produced, on the other. Our interdisciplinary study of Cupisnique kilns and associated ceramics suggests that early potters opted to use small, semi-closed kilns to gain control over the coloration of the ceramic surface and minimize breakage during firing. Most of these kilns had the capacity to fire at most a dozen bottles or perhaps two dozen bowls at a time. High local demand for ceramic vessels would force more frequent or simultaneous firing of multiple kilns by individual potters and lead to the observed spatial concentration of a large number of kilns.

The Cupisnique type semi-closed kiln persisted largely unchanged until about A.D. 300-350, a continuity of more than 1300 years. Unslipped, hand-formed late Gallinazo ceramics were found with the latest such kilns in the Poma Canal (Shimada et al. 1994). Nearly identical kilns were found in the Vicús region of the upper Piura Valley to the north where contemporaneous, hand-formed, unslipped Vicús ceramics predominated (Kaulicke 1991). This time period coincides with the spread of two major innnovations in pottery manufacture. The first few centuries A.D. on the north coast saw the establishment of a slip painting tradition, and increasing use of molds for vessel and figurine formation. The increasing use of slip-painted vessels bearing religious and political icons (particularly those rendered in fine-line) by the Mochica polity is believed to be positively related to these innovations. The development of new firing techniques and structures better suited for greater output and slip-painted ceramics -- e.g., increase in firing capacity and/or number of kilns that provided higher temperatures and consistent oxidizing atmosphere to intensify the color (hue) of slip paints and the number of vessels fired -- would thus be expected. This is a systemic response to changes brought to the integrated sequence of activities that we call pottery making. I believe the new kilns and firing techniques contributed to greater output of well-defined forms using molds, which was itself a reflection of greater sociopolitical complexity and control over information and material flow. Determination of the operational capabilities of partially enclosed rectangular kilns through replicative firing experiments and Mössbauer spectroscopy of kiln linings and ceramic products would be crucial in testing the inferences presented here.

Excavated pit kilns date back only to the late Mochica times (ca. A.D. 450-550), though their use probably began considerably earlier. They remain popular to this day on the Peruvian coast, surely due to their simple construction and operation. The pit kiln with some sort of refractory cover and lining reduces the loss of heat and affords even heat distribution and thus better fuel efficiency and more predictable results. Essentially, it creates a relatively stable, closed system. At the same time, it is critical to anticipate the necessary quantity and type of fuel and its distirbution within the kiln for optimal firing. Once ignited, the addition or removal of fuel would be difficult without producing some adverse effects. Large body sherds are the most common cover and linings found in the excavated pit kilns, cross-cutting time and space.

The quality of firing also depends to a surprising degree on how carefully vessels and fuel are arranged within the kiln, e.g., multi-tiered loading of different sized vessels and even distribution of fuel, versus indirect firing whereby vessels do not come into direct contact with fire. This and many other issues surrounding kilns and firing cannot be resolved simply by locating and excavating kilns but must be approached from a sustained, interdisciplinary perspective. These issues reflect a complex decision-making process affected not only by material constraints and potential but also by the interests and needs of the potters and their culture.

REFERENCES CITED

Anders, M.
1990 Maymi: Un Sitio del Horizonte Medio en el Valle de Pisco. *Gaceta Arqueológica Andina* 5(17):27-39. Lima.

Anders, M., V. Chang, and I. Shimada
1994 Taller de Cerámica Nazca Tardío-Huari en el Sitio de Maymi, Valle de Pisco. In *La Tecnología y la Organización de Producción de las Cerámicas en los Andes Prehispánicos*, edited by I. Shimada, pp. 249-267. Fondo Editorial, Pontificia Universidad Católica del Perú.
In press Early Middle Horizon Pottery Production at Maymi, Pisco Valley, Peru. In *Ceramic Production in the Prehispanic Andes: Technology, Organization, and Approaches,* edited by I. Shimada. Supplement to MASCA Research Papers, Vol. 14. MASCA, The University Museum, University of Pennsylvania, Philadelphia.

Bawden, G.
1977 Galindo and the Nature of the Middle Horizon on the Northern North Coast. Unpublished Ph. D. dissertation, Department of Anthropology, Harvard University, Cambridge, MA.

Carmichael, P. H.
1986 Nasca Pottery Construction", *Ñawpa Pacha* 24:31-48.

Cleland, K.M. and I. Shimada
1994 Cerámica Paleteada: Tecnología, Modos de Producción y Aspectos Sociales. In *La Tecnología y la Organización de Producción de las Cerámicas en los Andes Prehispánicos*, edited by I. Shimada, pp. 321-348. Fondo Editorial, Pontificia Universidad Católica del Perú.

Craig, A. K. and I. Shimada
1986 El Niño Flood Deposits of Batán Grande, Northern Peru. *Geoarchaeology* 1:29-38.

Donnan, C.B.
1965 Moche Ceramic Technology. *Ñawpa Pacha* 3:115-138.

Hayashida, Frances M.
1994 Producción Cerámica en el Imperio Inca: Una Vision Blobal y Nuevos Datos. In *LaTecnología y la Organización de Producción de las Cerámicas en los Andes Prehispánicos*, edited by I. Shimada, pp. 443-475. Fondo Editorial, Pontificia Universidad Católica del Perú, Lima.

Kaulicke, P.
1991 El Período Intermedio Temprano en el Alto Piura: Avances del Proyecto Arqueológico "Alto Piura" (1987-1990). *Bulletin de l'Institut Françcais d'Etudes Andines* 20:381-422. Lima.

Killick, D.
n.d. Petrographical Analysis of Peruvian Ceramics. Unpublished report prepared for Izumi Shimada. Archaeometry Laboratory, Department of Anthropology, Harvard University, Cambridge, MA.

Kroeber, A.L.
1930 *Archaeological Explorations in Peru, Part II: The Northern Coast.* Field Museum of Natural History, Anthropology Memoirs 2(2). Chicago.

Larco, R.
1941 *Los Cupisniques.* La Crónica y Variedades, S.A., Lima.
1948 Cronología Arqueológica del Norte del Perú. Sociedad Geográfica Americana, Buenos Aires.
Linné, S.
1925 *The Technique of South American Ceramics.* Göteborg Ethnological Museum, Göteborg.
Menzel, D., J. H. Rowe, and L. E. Dawson
1964 *The Pracas Pottery of Ica: A Study in Style and Time.* University of California Publications in American Archaeology and Ethnology, Vol. 50. Berkeley.
Moseley, M.E.
1975 Prehistoric Principles of Labor Organization in the Moche Valley, Peru. *American Antiquity* 40:141-196.
Pozzi-Escot, D., M. Alarcón, and C. Vivanco
1994 Cerámica Wari y su Tecnología de Producción: La Visión desde Ayacucho. In *La Tecnología y la Organización de Producción de las Cerámicas en los Andes Prehispánicos,* edited by I. Shimada, pp. 269-294. Fondo Editorial, Pontificia Universidad Católica del Perú, Lima.
Salazar, J., E. López, and J. N. Vreeland
1993 Medida de las Temperaturas de un Horno Artesanal de Mórrope, Departamento de Lambayeque, Utilizado para la Fabricación Tadicional de Alfarería. *Boletin del Instituto Francés de Estudios Andinos* 22(3):685-699.
Salazar, J., U. Wagner, F. E. Wagner, W. Korschinsky, M. Zahn, J. Riederer, and F. Kauffman-Doig
1986 Mössbauer Spectroscopy and Neutron Activation Analysis of Recent Finds from Chavin. In *Proceedings of the 1984 International Symposium on Archaeometry,* edited by J.S. Olin and M.J. Blackmann, pp. 143-151. Smithsonian Institution Press, Washington, D.C.
Shimada, I.
1978 Economy of Prehistoric Urban Context: Commodity and Labor Flow in Moche V Pampa Grande, Peru. *American Antiquity* 43:569-592.
1994 *Pampa Grande and the Mochica Culture.* University of Texas Press, Austin.
1995 Prehispanic Metallurgy and Mining in the Andes: Recent Advanes and Future Tasks. In *In Quest of Mineral Wealth: Aboriginal and Colonial Mining and Metallurgy in Spanish America*, edited by R. West and A. K. Craig, pp. 37-73. Geoscience and Man, Vol. 53. Louisiana State University, Baton Rouge.
Shimada, I. (editor)
1994 *La Tecnología y la Organización de Producción de las Cerámicas en los Andes Prehispánicos.* Fondo Editorial, Pontificia Universidad Católica del Perú, Lima.
In press *Ceramic Production in the Prehispanic Andes: Technology, Organization, and Approaches.* Supplement to MASCA Research Papers, Vol. 14. MASCA, The University Museum, University of Pennsylvania, Philadelphia.

Shimada, I., V. Chang, and C. Elera
1990 Excavaciones en Hornos de Cerámica de la Epoca Formativa en Batán Grande, Costa Norte del Perú. *Gaceta Arqueológica Andina* 5(20):19-43. Lima.

Shimada, I., V. Chang, C.G. Elera, H. Neff, M. Glascock, U. Wagner, and R. Gebhard
1994 Los Hornos y la Producción de Cerámica Formativa en Batán Grande, Costa Norte del Perú. In *La Tecnología y la Organización de Producción de las Cerámicas en los Andes Prehispánicos*, edited by I. Shimada, pp. 67-119. Fondo Editorial, Pontificia Universidad Católica del Perú, Lima.

Shimada, I. and J.F. Merkel
1991 Copper Alloy Metallurgy in Ancient Peru. *Scientific American* 265(1):80-86.

Shimada, M. and I. Shimada
1985 Prehistoric Llama Breeding and Herding on the North Coast of Peru. *American Antiquity* 50:3-26.

Stone-Miller, R.
1995 *Art of the Andes from Chavín to Inca*. Thames and Hudson, London.

Tello, J.C.
1938 Arte Antiguo Peruano, Primera Parte: Tecnología y Morfología. *Inca* 2: vii-lxii, 1-280. Lima.

Topic, T.L.
1977 *Excavations at Moche*. Unpublished Ph. D. dissertation, Department of Anthroplogy, Harvard University, Cambridge.

Tschauner, H., M. Vetters, J. Dulanto, M. Saco, and C. Wester
1994 Un Taller Alfarero Chimú en el Valle de Lambayeque. In *La Tecnología y la Organización de Producción de las Cerámicas en los Andes Prehispánicos*, edited by I. Shimada, pp. 349-393. Fondo Editorial, Pontificia Universidad Católica del Perú, Lima.

Uceda, S. and J. Armas
In press Estudio de un Taller de Alfarero en el Centro Urbano de Moche, Valle de Moche. In *Ceramic Production in the Prehispanic Andes: Technology, Organization, and Approaches,* edited by I. Shimada. Supplement to MASCA Research Papers, Vol. 14. MASCA, The University Museum, University of Pennsylvania, Philadelphia.

Uhle, M.
1913 Die Ruinen von Moche. *Journal de la Société des Américanistes de Paris* n.s. 10:95-117.
n.d. Final Report on the Excavations at the Site of Moche in 1899 and 1900. Manuscript deposited in the Ibero-Amerikanisches Institut, Berlin.

Wagner, U., R. Gebhard, E. Murad, G. Grosse, J. Riederer, I. Shimada, and F.E. Wagner
n.d. Formative Ceramics from the Andes and their Production: A Mössbauer Study. Paper presented at the Fifth Latin American Conference on Applications of the Mössbauer Effect, Sept. 9-14, 1996, Cusco. To be published in *Hyperfine Interactions.*

Wagner, U., R. Gebhard, E. Murad, J. Riederer, I. Shimada, C. Ulbert, and F.E. Wagner
In press Production of Formative Ceramics: Assessment by Physical Methods. In *Ceramic Production in the Prehispanic Andes: Technology, Organization, and Approaches,* edited by I. Shimada. Supplement to MASCA Research Papers, Vol. 14. MASCA, The University Museum, University of Pennsylvania, Philadelphia.

Wagner, U., R. Gebhard, E. Murad, J. Riederer, I. Shimada, C. Ulbert, F.E. Wagner, and A. M. Wippern
1994a Las Condiciones de Quemas y Otros Aspectos Tecnológicos de las Cerámicas Formativas de los Andes: Perspectiva Arqueométrica. In *La Tecnología y la Organización deProducción de las Cerámicas en los Andes Prehispánicos*, edited by I. Shimada, pp. 121-156. Fondo Editorial, Pontificia Universidad Católica del Perú, Lima.

Wagner, U., R. Gebhard, E. Murad, J. Riederer, J., I. Shimada and F.E. Wagner
1994b Kiln Firing at Batán Grande: Today and in Formative Times. In *Archaeometry of Pre-Columbian Sites and Artifacts,* edited by D. A. Scott and P. Meyers, pp. 67-84. The Getty Conservation Institute, Marina del Rey, CA.

Willey, G. R.
1949 Ceramics. In *Handbook of South American Indians, Vol. 5: The Comparative Ethnology of South American Indians,* edited by J. H. Steward, pp. 139-204. Smithsonain Institution, Bureau of American Ethnology, Bulletin 143. Washington, D.C.
1964 Diagram of a Pottery Tradition. In *Process and Pattern in Culture: Essays in Honor of Julian H. Steward,* edited by R. A. Manners, pp. 156-174. Aldine Publishing Co., Chicago.

Wilson, D.L.
1988 *Prehispanic Settlement Patterns in the Lower Santa Valley, Peru: A Regional Perspective on the Origins and Development of Complex North Coast Society*. Smithsonian Institution Press, Washington, D.C.

CERAMIC FIRING IN ANCIENT AND MODERN OAXACA

Gary M. Feinman and Andrew Balkansky, University of Wisconsin-Madison, Department of Anthropology, 1180 Observatory Drive, Madison, WI 53706

ABSTRACT

In the Valley of Oaxaca, Mexico, ceramic firing practices were variable in prehispanic times, as they are today. These techniques are described, including a recently excavated pit kiln from the Classic period Ejutla site. Based on an overview of firing technology in Oaxaca, a simple linear trajectory from open firing to pit kilns to updraft kilns is not apparent through time, nor do these technological differences seem simply to reflect discrete ethnic patterns or population replacements. Likewise, the diversity in firing practices cannot be accounted for by the scalar distinction between household and factory production. Rather, past and present variation in firing technology appears most explicable as a function of the volume or intensity of pottery manufacture.

INTRODUCTION

Variability in methods of ceramic production are characteristic of present-day Mesoamerica. The renowned ethnologist George M. Foster (1967:103) observed: "For variety in pottery-making techniques and in materials, forms, styles, and economic importance, few places in the world rival Mesoamerica and, particularly, Mexico" (Figure 1). Foster attributed the variation in modern ceramic production to ethnicity, arguing that distinct social groups made pottery in diverse ways, according to their unique cultural institutions and histories. According to Foster (1955, 1965, 1967), the technological heterogeneity of modern potting communities had become especially pronounced with the introduction of the European updraft kiln into mestizo towns after the Spanish conquest. Yet in a departure from the trait-based framework of culture history, Anna O. Shepard (1963), noting the variation in contemporary methods of production and patterns of marketing in the Oaxaca Valley, argued that some of the described variability was functionally related to socioeconomic differences between pottery-making communities. Shepard added, moreover, that valuable economic information could be discerned by archaeologists from the study of ancient processes of ceramic manufacture and distribution.

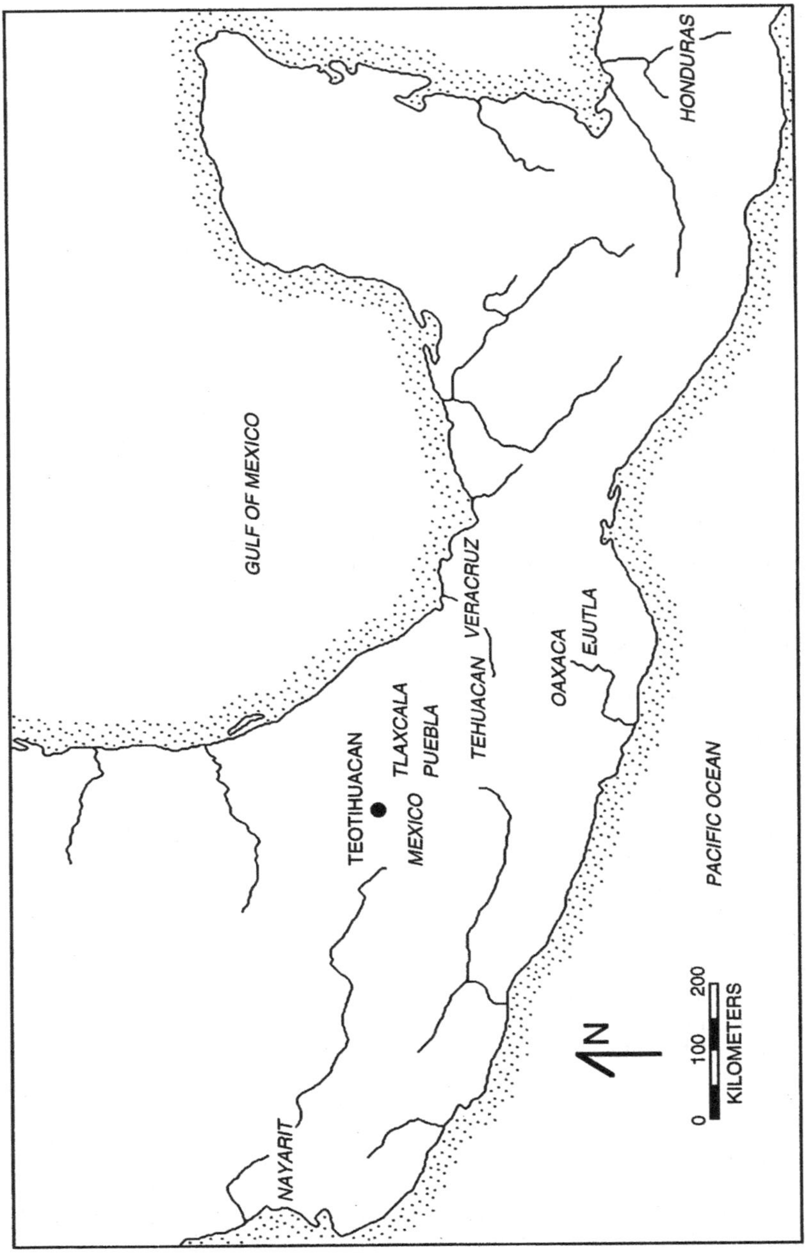

Figure 1. Southern Mexico, showing places mentioned in the text

This ethnographic perspective on ceramic production and firing presents an intriguing series of questions for the archaeologist. What was the nature of variability in the past? Did diversity come about only with European contact? If variation existed in prehispanic times, what factors might have accounted for it? How did the nature of this variation shift over time? How does the prehispanic Oaxaca case inform us about ceramic manufacturing technology? The recent excavation of ancient pottery kilns at the Ejutla site in Oaxaca (Mexico) (Figure 2) provides an avenue to address these questions of prehistoric variability in methods of ceramic production, especially if we consider the Ejutla features in light of other prehispanic Mesoamerican pottery production. Before presenting the archaeological findings from Ejutla, we further examine traditional pottery-making in Oaxaca.

MODERN POTTERY VILLAGES IN OAXACA

Comparisons among the Oaxaca Valley's traditional pottery-producing villages illustrate much of the diversity cited by Foster, with differences in the intensity of production, choice of firing methods, and patterns of distribution (see overviews in Flannery and Marcus 1994; Shepard 1963). Our use of the word "intensity" here follows Costin (1991), who distinguishes the amount of time devoted to a craft and the volume of goods produced (intensity of production) from another dimension, the scale of production. The latter refers to the specific social situation of the producers (household, workshop, or factory).

The villages of Santa María Atzompa, San Bartolo Coyotepec, San Marcos Tlapazola, and Ocotlán de Morelos (see Figure 2) have received uneven ethnographic attention (e.g., Foster 1955; Hendry 1992; Stolmaker 1976; van de Velde and van de Velde 1939). Each of these communities is characterized by a fairly low, though variable, intensity of household-based ceramic production. Atzompa, located less than 5 km from the region's largest market in the city of Oaxaca de Juárez, represents one end of the production intensity continuum. The Atzompa potters make large numbers of distinctive green-glazed ornamental vessels and figurines that are destined for regional markets in Oaxaca City and commercial venues far beyond. Their pottery is fired in cylindrical updraft kilns, in which the fuel is placed in a shallow firebox excavated into the ground surface below the vessel chamber. The Atzompa kilns are relatively permanent facilities made from stone or adobe, mortar, and reused material from older kilns. These kilns are broadly analogous in design to those known from ancient Monte Albán (Winter and Payne 1976).

In contrast, the contemporary potters at still remote San Marcos do not use a formal kiln, and instead fire their far smaller and less diverse inventory of plain brown wares with an open or bonfire method. The San Marcos potters market most of their tortilla griddles (*comales*) and other utilitarian bowls and jars in the closest regional market in Tlacolula (and to a lesser degree in Oaxaca City). In San Marcos, there are no permanent firing facilities, the pottery and fuel are intermingled on the ground surface and burned in the open air. This relatively impermanent firing method leaves few definitive traces for the archaeologist

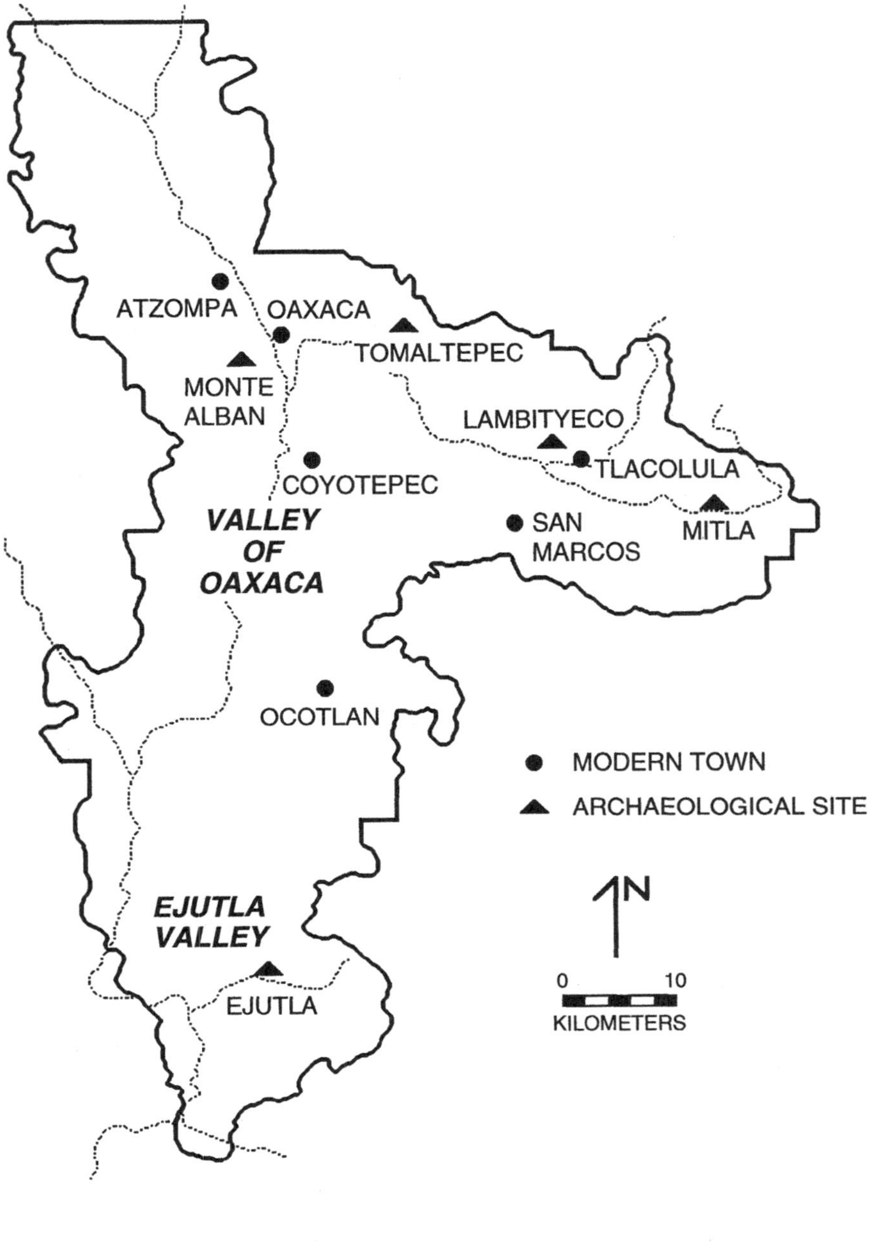

Figure 2. The Oaxaca and Ejutla valleys, with places mentioned in the text

(Payne 1970, 1994). It is provocative that the respective contexts of production at Atzompa and San Marcos are so similar--occurring in and around family houselots--despite such marked differences in firing methods, intensity of production, and spheres of distribution. Coyotepec and Ocotlán potters also operate primarily in household contexts, but producers in these communities fall between Atzompa and San Marcos in terms of the volume of manufacture and the distances over which their wares are distributed.

ARCHAEOLOGICAL FIRING FEATURES

The discovery of updraft kilns at Monte Albán (Winter and Payne 1976) provided evidence that the more formal firing technology seen at Atzompa had antecedents in the prehispanic era (also see Payne 1982; Santley, Arnold, and Pool 1989; Shimada 1989; Winter 1994:21; Winter and Nardin 1982). In Mesoamerica, true kiln firing was not a product of European contact. The Monte Albán kilns were built of stone and mortar, with firing chambers excavated into the bedrock and ware chambers formed over a grill of adobe blocks. These substantial features were associated with domestic residences and dated to the Classic period (A.D. 200-800) (Table I).

Table I. Chronological Sequences

	VALLEY OF OAXACA	MESOAMERICA
1500 AD		
	MONTE ALBAN V	LATE POSTCLASSIC
1200		
900	MONTE ALBAN IV	EARLY POSTCLASSIC
600	MONTE ALBAN IIIB	LATE CLASSIC
	MONTE ALBAN IIIA	EARLY CLASSIC
300		
0	MONTE ALBAN II	TERMINAL FORMATIVE
300	MONTE ALBAN LATE I	LATE FORMATIVE
	MONTE ALBAN EARLY I	
600	ROSARIO	MIDDLE FORMATIVE
	GUADALUPE	
900		
	SAN JOSE	
1200		EARLY FORMATIVE
	TIERRAS LARGAS	
1500 BC		

A variant (but apparently less substantial) type of indirect-firing kiln was found in a cave occupied during the Late Postclassic period (A.D. 950-1520) near Mitla (Flannery and Marcus 1983:299). This feature was used to fire a large number of gray-ware jars that were densely deposited near the kiln. This kiln consisted of an adjacent pair of circular chambers separated by a stone-lined tunnel or flue. This firing arrangement might be likened to an "horizontal" updraft kiln. Three much earlier two-chambered features (in which the chambers were separated by an adobe wall) were unearthed west of Mitla at the Tomaltepec site (see Whalen 1981:97). These subterranean Tomaltepec features also were found in household contexts, but dated to Monte Albán Late I (300-200 B.C.). The majority of these features were 60-80 cm in diameter. In such features, evenness in firing could have been achieved by placing the vessels in one chamber and the fuel in another (Flannery and Marcus 1996:170). Unfortunately, the specific artifacts associated with these features have not been described or reported. A more ephemeral firing feature was identified at Postclassic (A.D. 800-1520) Lambityeco (Swezey 1975) by the association of ash with heavy densities of pottery. This last instance may prefigure the modern firing methods of San Marcos (Payne 1970, 1994).

The aforementioned cases notwithstanding, most discussions of ceramic production in ancient Mesoamerica have had to rely on indirect measures. Pottery kilns--the most direct evidence of the production context--are curiously rare (e.g., Payne 1970; Santley, Arnold, and Pool 1989; Stark 1985). This empirical reality is perplexing, given the ubiquity of pottery in Mesoamerica, and the level of scholarly interest in the subject. Among other reasons given for the difficulty of locating firing areas is their distance from structures, the reuse of building material, and the ravages of time. In the absence of recognizable kilns, Mesoamerican archaeologists have relied on other material correlates of pottery production: principally, anomalous densities of potsherds and specific ceramic types, clay concretions, ash lenses, and wasters (e.g., Feinman 1980, 1982; Krotser 1987; Redmond 1979; Stark 1985). These criteria, however, are subject to equifinality, such that each class of data cannot be taken alone as a definitive indicator of ceramic production. To present a convincing demonstration of ancient production, "a pattern of evidence must be examined" (Stark 1989).

The Ejutla Pit Kilns

Excavations at Ejutla (1990-1993) uncovered the stone foundation of a Classic period house, along with evidence for ceramic and other craft production (Feinman and Nicholas 1995; Feinman, Nicholas, and Middleton 1993) (Figure 3). Dense midden deposits with an abundance of figurines and figurine wasters, figurine molds, and hundreds of wasters from a variety of ceramic vessels were found to overlie shallow features that the prehispanic inhabitants had dug into the soft bedrock. Despite this evidence, the initial implication that these bedrock features were substantially constructed pottery kilns did not fit our initial expectations, either ethnographic--based on the familiar production methods of the potters at Atzompa, or archaeological--as reported from the excavations of the more formal features at Monte Albán. In Ejutla, no stone or adobe walls were present.

We sought to develop expectations for the production of ceramics from a broad interpretive context, which included the bedrock features themselves, their

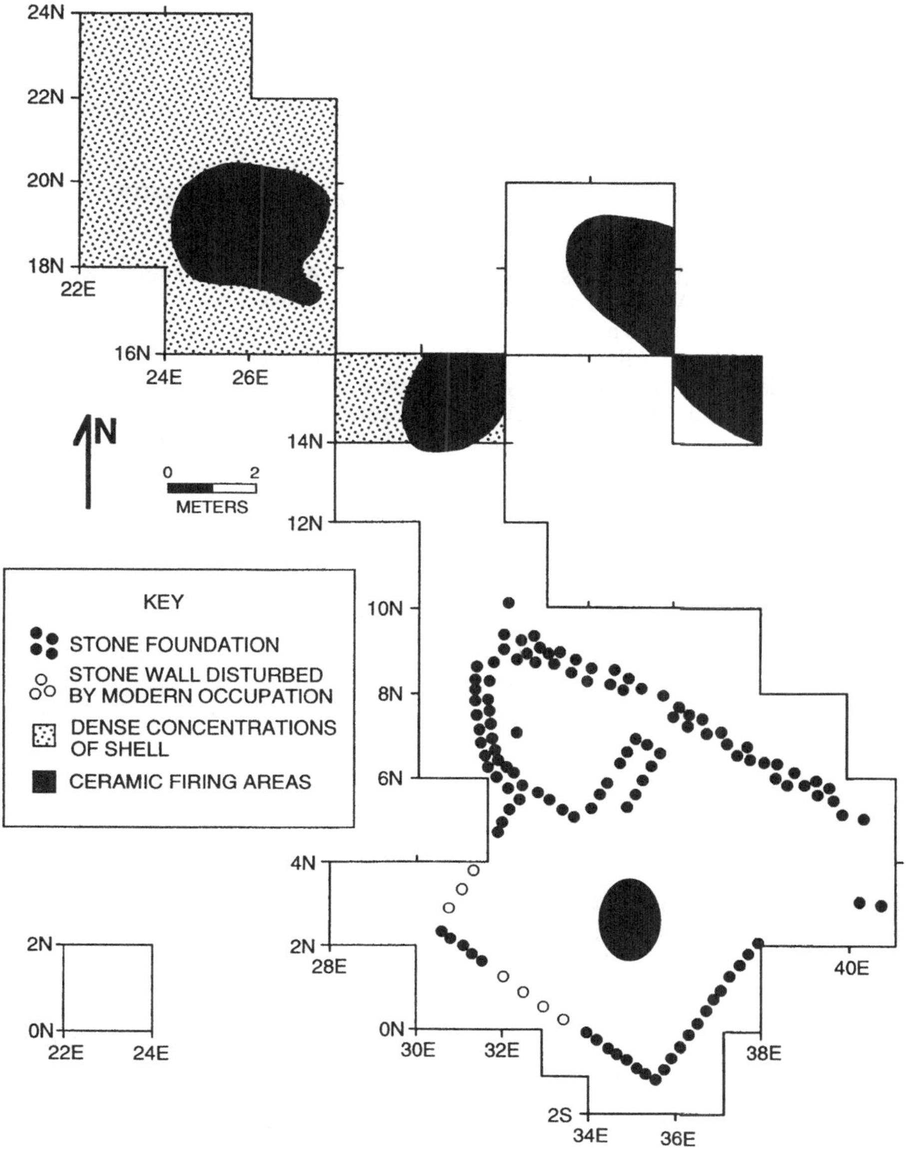

Figure 3. Area of excavation at the Ejutla site, illustrating the residential structure, firing features, and dense midden area

associated ceramic products and by-products, the results of a ceramic firing experiment, and extant data on informal methods of production reported from other prehispanic sites in Mesoamerica. In addition, the Ejutla discoveries provided an opportunity to record both the features used for ceramic firing and the associated artifacts--something that was not available for earlier prehispanic ceramic production facilities found in the Valley of Oaxaca. Based on our multi-faceted investigations, we suggest that ceramic firing at Ejutla was carried out in pit kilns.

In Ejutla, no formal kiln features like those at modern Atzompa or ancient Monte Albán were found. Although some large cobbles were associated with most of the pit features, there were not nearly enough stones to constitute a kiln structure (as observed at Monte Albán). Nevertheless, in Ejutla, the recovery of large numbers of defective sherds, pottery by-products, and the frequent disposal of broken (but unused and complete) ceramic forms suggested that the most abundant ceramic varieties--especially molded figurines, incense burners (*sahumadores*), tortilla griddles (*comales*), and certain types of reduced (gray) bowls--likely were made and fired in close proximity to the excavated structure. The distributions and densities of these particular forms distinguish them from all other ceramic types found at the site. The occurrence of these and other production-related ceramic artifacts, moreover, is localized in and adjacent to the bedrock features. The 2 x 2 m excavation squares that include the bedrock depressions had high densities of the aforementioned ceramic artifacts, deep ash layers, as well as concentrations of wasters and clay concretions, compared to other units that were excavated. The unearthing of at least four such bedrock features led us to suspect that these features were associated with the pottery-manufacturing process.

Most of Ejutla's pit kilns were overlain with dense midden deposits that contained a wide variety of artifacts. Even so, the levels below the midden deposits that were in direct association with the bedrock pits held a more restricted artifact complex. These deposits were limited to ash, ceramic wasters, clay concretions, stone cobbles, burned bedrock, and potsherds--all consistent with a ceramic firing context (Figure 4). The basal levels of these units held almost pure ash, often mixed with several large potsherds (often large upside-down jars) that may have been used to insulate the vessel chamber from the bedrock surface, or served as spacers, or other types of kiln furniture (Figure 5). In several of the pit features, we recovered nearly complete and unused vessels that were slightly flawed. These complete but slightly imperfect vessels appear to have been immediate discards after firing. Winter and Payne (1976:39) recovered similar complete vessels associated with the Monte Albán kiln. A sizable proportion of the ceramics found at Ejutla (and particularly those found near the pit features) were fire-clouded, multi-toned or completely misfired. These artifacts may indicate that the Ejutla artisans did not have complete control of the firing process.

At Ejutla, the surficial bedrock of the pit-firing features was blackened as if from burning, while decomposing pieces of bedrock that also had been charred were found in association. The best preserved of the pit-firing features (at 18N24E and

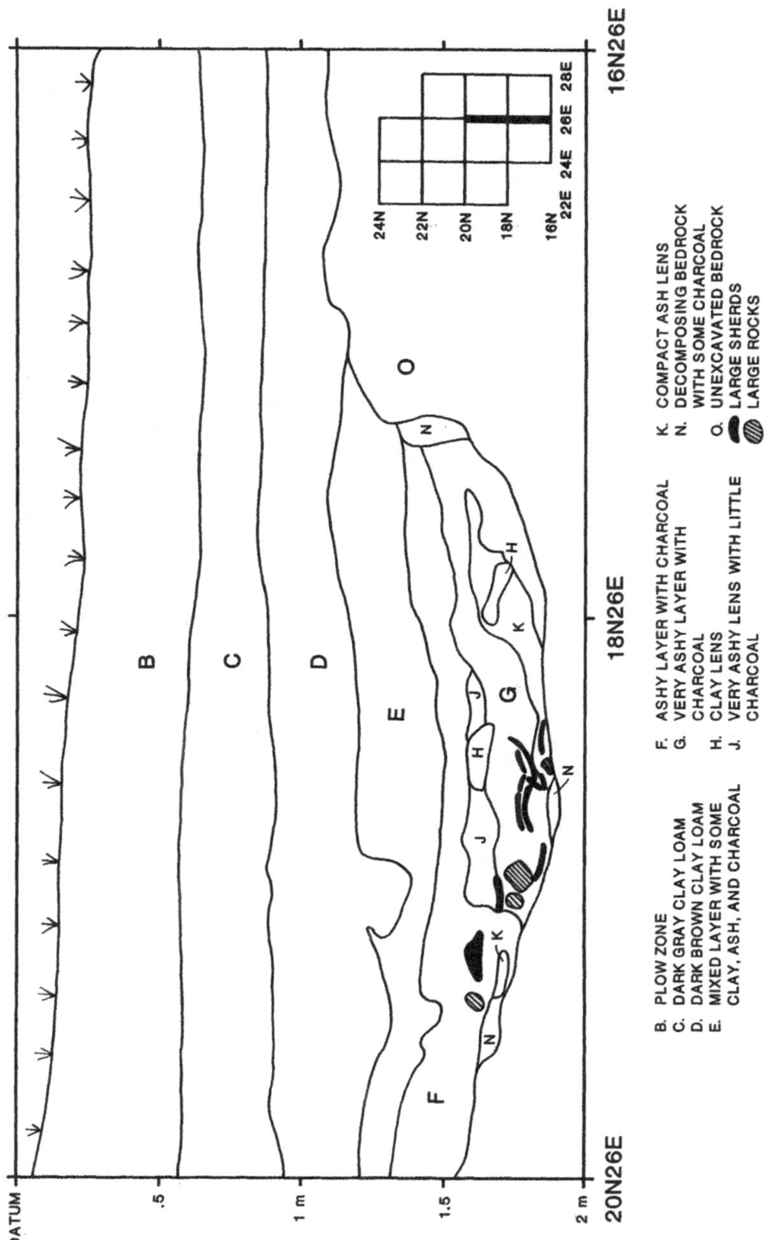

Figure 4. Profile of prehispanic pit kiln (18N24E), with mass of potsherds and ash layers at the base

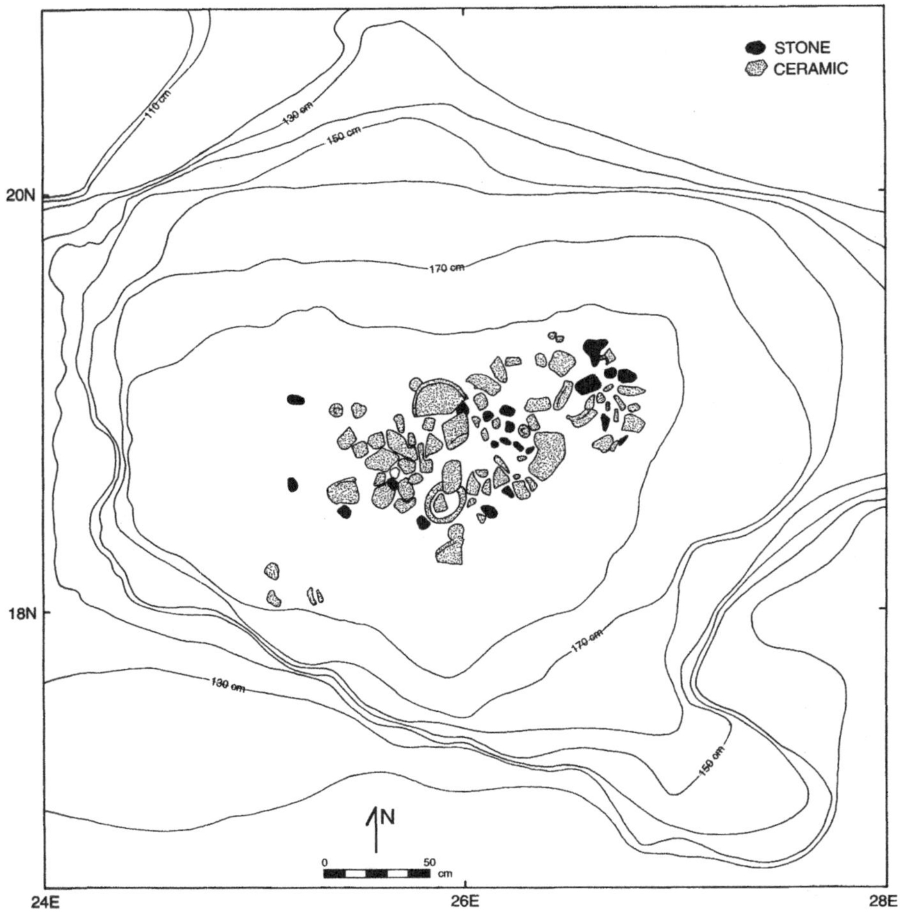

Figure 5. Contour drawing of the pit kiln in 18N24E, illustrating kiln furniture and other debris near the base

2N34E, see Figure 3) were slightly asymmetrical, with dimensions ranging from 2 to 4 m across and 40 to 70 cm deep. The irregular shape of these pit kilns was due in part to a narrowing of the bedrock depression into what might have served as a stoke pit, or mouth, to add fuel and ventilate the kiln. Each of the extensions was oriented to the south--the direction of the prevailing afternoon wind--which presumably enhanced performance with increased airflow into the kiln. Similar orientations toward local prevailing winds have been reported ethnographically (e.g., Krotser 1980) and noted at other archaeological sites (Abascal 1975; Bordaz 1964; Rice 1994; Russell 1994). In at least two instances, the bedrock appeared to have been modified after its initial use: the feature in excavation unit 14N30E was dug to a second, deeper level, while the "cigar-shaped" feature in units 16N34E and 14N36E may represent two separate, partially overlapping kilns. Despite such minor variations, both the nature of these features and their associated artifactual remains suggest that each served as a relatively impermanent means to pit-fire pottery.

EXPERIMENTAL FIRING

To broaden the interpretive perspective for the archaeological firing features at Ejutla, an experimental pit kiln was constructed and fired at the outdoor archaeology facility at the University of Wisconsin. Since there is no contemporary analog to pit-firing in the Oaxaca Valley, the design of the experimental pit kiln was modeled after ethnographic descriptions of traditional potters in India, Pakistan, and the southwestern United States (e.g., Rice 1987; Rye 1981; Shepard 1961; Sinopoli 1991). We recognize that pit kilns are not, in a literal sense (see Rhodes 1968:11), "true ceramic kilns" in that they do not provide complete separation between the fuel and the ceramic vessels being fired. Our use of the term "pit kiln," although derived from the literature of pottery-making in South Asia (Rye 1981; Rye and Evans 1976; Sinopoli 1991), has been applied more broadly as well (e.g., Heacock; Rice 1987:158).

For the experiment, a shallow depression was dug into the ground surface, and its contours defined by a layer of sand as a referent for the post-firing excavation. The chamber was filled with a mixture of pottery (in this instance, various vessel forms, including comales and test tiles), kiln furniture (to separate the vessels and improve airflow), and fuel (cow dung, grasses, and fallen tree branches). The mixture of pottery and fuel was then covered with straw to form a matrix upon which to construct a mud plaster roof. The mud plaster provides structure--to avoid the damage wrought by shifting of the pots as the fuel is consumed--and insulation--to conserve fuel and better control temperature and atmosphere. The atmospheric conditions can be modified throughout the firing by opening air holes in the mud plaster dome, or, alternatively, by smothering the fire to create a reducing atmosphere. The enclosed nature of these kilns, compared to other firing methods, allows potters a ready means to smother the fire (Rye 1981:98). In the past, pit kilns would have been especially well suited to producing the reduced gray-ware vessels prominent in the prehispanic tradition of the Oaxaca Valley (Caso and Bernal 1965).

The experiment provided an independent perspective on the spatial patterns of the material remains, the condition of the underlying soil, firing temperatures, and other characteristics designed to examine the feasibility of a pit-firing model for the Ejutla features. The excavated experimental kiln section left few indications that a stage in ceramic manufacturing had occurred there. Evidence was limited to a number of broken vessels that had been damaged in the firing, a thin layer of ash mixed with charcoal, and a ground surface beneath the kiln that was discolored and partially baked to a depth of 1 cm. Few of these residual materials from the experiment would likely survive intact on an exposed surface. On the other hand, if these deposits were rapidly sealed, they may have remained detectable as unusual concentrations of potsherds--perhaps particular ceramic types--and wasters. Most of the firing features at Ejutla were covered by subsequent midden accumulation (or in the case of the southernmost Ejutla feature, by a house floor). The most prominent remnant of the experimental firing was the pit kiln's mud plaster covering, now fired and broken, that strongly resembled the clay concretions abundant at Ejutla and other pottery-making sites in Mesoamerica (e.g., Bordaz 1964; Redmond 1979; Stark 1985). Although other processes might result in similar concretion-like materials, their presence in contexts where numerous other indicators of ceramic production (including the pit-kilns) have been recovered indicates that these artifacts may be all that remain from impermanent roofing. The implication of the experimental firing is that ancient Mesoamerican potters could have used a pit-firing technology for the specialized production of ceramics, while leaving relatively few traces of the ceramic-firing process.

IMPLICATIONS AND DISCUSSION

A final line of inference was drawn from a review of the ceramic-firing literature. A number of cases similar to those described ethnographically, and consistent with the material remains at Ejutla, suggest that impermanent firing features may be more common than generally supposed. Relatively impermanent kilns have been noted at Lambityeco (Swezey 1975) and Teotihuacan (Sheehy 1992), as well as at the Peñitas site in coastal Nayarit (Bordaz 1964), and several sites in the Puebla-Tlaxcala region (Abascal 1975). The recent discovery of pit kilns in Peru (Russell 1994; see also Blinman, this volume; Shimada, this volume) suggests that these more ephemeral features for ceramic firing may have been more common in other New World regions as well.

Ceramic production at Ejutla was clearly a household-scale enterprise. This is not surprising since careful archaeological investigations in Mesoamerica repeatedly situate craft production in a domestic context (e.g., Charlton, Nichols, and Charlton 1991; Manzanilla 1993). But what of the intensity of that production? With the emphasis on shell ornament and other craft production, the residents of this Ejutla household could not have been full-time potters. Yet the sheer volume of potsherds, and especially the focus on a few specific types, indicates that at least some of this pottery was intended for extra-household consumption. While it is difficult to quantify the relative intensity of ceramic production at Ejutla compared to other Mesoamerican households, it may be possible to gain some comparative insight through the analysis of the firing technology.

In a cross-cultural review of 55 modern potting communities, Sheehy (1992:73-102) compares resources, social organization, technology and other aspects of production to construct a multi-level model of ceramic production intensity. Although Sheehy recognized the difficulty of drawing generalizations from such a wide-ranging sample, he noted a proportionate increase between the frequency of ceramic firing and the formality of the firing facilities (1992:133-140). The lowest-volume producers fire their pottery in open hearths or pit kilns. Levels of productive intensity increase with the use of updraft kilns and still more complex down-draft and multi-chambered kiln arrangements (changes that Sheehy attributes to demand). Even within the category of pit-firing groups, there is a trend toward increasing diameter and depth of the firing facility with greater levels of production. The strong correlation between the intensity of ceramic production and the formality of firing technology identified by Sheehy provides a well-founded cross-cultural basis for the proposition that the choice of firing technology is intensity-related. In our comparison of ceramic production in Oaxaca, it is revealing that, although potters at both Monte Albán and Ejutla worked in similar domestic contexts, the firing technologies they employed were quite different. Compared to Ejutla, ceramic manufacturing at Monte Albán may have been conducted more continuously and at greater intensities, making the construction and maintenance of more formal and permanent firing facilities a reasonable if not necessary investment.

We are unaware of any contemporary Mesoamerican examples of pit-kiln firing. Perhaps for this reason, Mesoamerican archaeologists have not consistently recognized informal firing features in the vain expectation that more formal updraft kilns were generally employed. Yet the findings presented here suggest that in ancient Mesoamerica significant quantities of pottery were fired in pit kilns and other less permanent facilities. Ephemeral means of ceramic firing likely were employed at various ancient Mesoamerican sites where ample evidence for ceramic production has been noted, but no formal kilns have been recovered (e.g., Krotser 1987; Rattray 1990). Mesoamerican archaeologists should no longer be predisposed to encounter formal updraft kilns alone. The problem of identifying prehispanic ceramic production is not simply a product of the "tyranny of ethnographic analogy," but may be compounded by the implementation of field methodologies that are not always adequate to recover impermanent exterior features. Excavation strategies limited to narrow test trenches or isolated 2 x 2 m units are unlikely to yield unequivocal archaeological indicators of ephemeral firing. Inattention to midden deposits undoubtedly also has led archaeologists to define firing features that were later filled with trash as "burned areas within middens," or the ubiquitous "ash-filled pit." Without careful lateral excavations of exterior contexts that immediately surround non-elite domestic residences, the systematic recognition of impermanent firing features will remain elusive.

Despite these methodological limitations, extant findings indicate that diverse firing strategies have been practiced in Mesoamerica since the Formative period.

The known distribution of relatively permanent updraft kilns (like those at Monte Albán) varies widely over time--from the Late-Terminal Formative (300 B.C.-A.D. 200) through the Late Postclassic periods--and across space--in Oaxaca, Tehuacán, the state of Mexico, Veracruz, and Honduras (Cabrera Castro 1988; Flannery and Marcus 1983:299; Redmond 1979; Santley, Arnold, and Pool 1989; Stone and Turnbull 1941; Whalen 1981:97; Winter 1994; Winter and Payne 1976). This distribution is matched by the reported sample of impermanent, open, or pit-firing facilities from a remarkably diverse range of sites: village occupations in Middle and Late Formative Puebla-Tlaxcala (Abascal 1975, 1976), Classic-period Teotihuacan (Sheehy 1992), and Postclassic sites in Nayarit (Bordaz 1964) and Oaxaca (Swezey 1975). The Classic-period Ejutla pit kilns apparently were used at the same time as the more formal kilns at Monte Albán, but varied in size, design, and degree of permanence. In highland Oaxaca, therefore, technologically diverse ceramic-firing methods were practiced at contemporaneous sites less than 75 km apart.

If we look at all of the known Valley of Oaxaca and neighboring Ejutla firing features, it also becomes clear that there was no simple, linear trend toward more substantial or more formal firing features in the region. Two chambered features appear to have been constructed as early as the Formative period at Tomaltepec (Whalen 1981). By the Terminal Formative period (200 B.C.-A.D. 200), fairly substantial firing facilities were apparently erected at Monte Albán (although basic descriptive data remain unpublished). Substantial kilns also were used at that same urban site during the Classic period (Winter and Payne 1976). Yet the least permanent prehispanic firing features known for Oaxaca date to the same later Classic period (Ejutla) as well as subsequent the Postclassic (Lambityeco). Although we have few examples of Postclassic period ceramic firing in Oaxaca, we note diversity here also between the Lambityeco findings and the somewhat more substantial firing features in Guilá Naquitz cave near Mitla.

CONCLUSION

The research presented here suggests that the modern diversity of the potters' craft described by Foster has antecedents in the prehispanic era. But how might we account for these variable production methods over so long a time span? If ethnicity were the key factor, one might expect differences between informal pit-firing and formal updraft kiln sites in vessel forms and decorative attributes, as well as method of production. Contrary to these expectations, the ceramic complex from Ejutla was highly similar on stylistic grounds to Classic period occupations in the Oaxaca Valley. If ethnicity explained the recognized differences in firing technologies, then clearer geographic and temporal distinctions also might be supposed. Yet this view is not supported by the cases cited here, which clearly do not vary along a single spatio-temporal trajectory.

We also question whether ceramic technology can always be seen as advancing in a regular of linear manner. That is once formal kilns were constructed by prehispanic Oaxacan potters, these features were not uniformly used by ceramic producers throughout the region. Like the modern potters of the region, the ancient pottery producers adjusted their technology in accordance with a range of considerations. We adhere to an approach closer to that advocated by Shepard, one that seeks an active relationship between the organization of ceramic producers, their technology, and multi-scale patterns of consumption. A full accounting of the past variation and change in ancient Mesoamerica's ceramic production complex must include socioeconomic and demographic factors in the analysis.

To conclude, we have argued that Mesoamerica's ethnographic record, while an indispensable source of archaeological insight, cannot be applied uncritically to the prehispanic past of the region. There appears to have been greater variability in ceramic-firing practices in ancient than in modern times. Archaeological methods may play a role in the under-representation of this variability, and especially in the identification of less intense modes of ceramic production. What is needed are more excavated production contexts that are not limited to isolated features, but include households and the lateral space beyond structures. We also have proposed that differences in the intensity of ceramic production might be reflected in the choice of firing technology. These technological choices are often more indicative of extant socioeconomic conditions than ethnicity, progressive forces of history, or simple unilinear models of evolution. Finally, the research at Ejutla suggests that in ancient Mesoamerica, ceramic production was often effected by the deceptively simple technology of the pit kiln.

References

Abascal, R.

1975 Los Hornos Prehispánicos en la Región de Tlaxcala (Prehispanic Kilns in the Tlaxcala Region). *XIII Mesa Redonda de la Sociedad Mexicana de Antropología*, tomo I: 189-198. Mexico.

1976 Los Primeros Pueblos Alfareros Prehispánicos (The First Prehispanic Ceramic-Making Villages). In *Comunicaciones, Suplemento, vol. III, Proyecto Puebla-Tlaxcala*, pp. 49-52. Fundación Alemana para la Investigación Científica.

Bordaz, J.

1964 *Pre-Columbian Ceramic Kilns at Peñitas, a Post-Classic Site in Coastal Nayarit, Mexico*. Unpublished Ph.D. dissertation, Columbia University, New York.

Cabrera Castro, R.

1988 Horno Cerámico Posteotihuacano en el Palacio de Atetelco (Post-Teotihuacan Ceramic Kiln in the Atetelco Palace). *Arqueología* 4: 47-75.

Caso, A., and I. Bernal.

1965 Ceramics of Oaxaca. In *Handbook of Middle American Indians, vol. 3: Archaeology of Southern Mesoamerica, part 2*, edited by R. Wauchope and G. R. Willey, pp. 871-895. University of Texas Press, Austin.

Charlton, T. H., D. L. Nichols, and C. O. Charlton.

1991 Aztec Craft Production and Specialization: Archaeological Evidence from the City-State of Otumba, Mexico. *World Archaeology* 23: 98-114.

Costin, C. L.
1991 Craft Specialization: Issues in Defining, Documenting, and Explaining the Organization of Production. In *Archaeological Method and Theory*, vol. 3, edited by M. B. Schiffer, p. 1-56. University of Arizona Press, Tucson.

Feinman, G. M.
1980 *The Relationship Between Administrative Organization and Ceramic Production in the Valley of Oaxaca, Mexico*. Unpublished Ph.D. dissertation, City University of New York, New York.
1982 Patterns in Ceramic Production and Distribution, Periods Early I Through V. In *Monte Albán's Hinterland, Part 1: Prehispanic Settlement Patterns of the Central and Southern Parts of the Valley of Oaxaca, Mexico*, pp. 181-206. Memoir 15. Museum of Anthropology, University of Michigan, Ann Arbor.

Feinman, G. M., and L. M. Nicholas.
1995 Household Craft Specialization and Shell Ornament Manufacture in Ejutla, Mexico. *Expedition* 37: 14-25.

Feinman, G. M., L. M. Nicholas, and W. D. Middleton.
1993 Craft Activities at the Prehispanic Ejutla Site, Oaxaca, Mexico. *Mexicon* 15: 33-41.

Flannery, K. V., and J. Marcus.
1983 Urban Mitla and its Rural Hinterland. In *The Cloud People: Divergent Evolution of the Zapotec and Mixtec Civilizations*, edited by K. V. Flannery and J. Marcus, pp. 295-300. Academic Press, New York.
1994 *Early Formative Pottery of the Valley of Oaxaca, Mexico*. Memoir 27. Museum of Anthropology, University of Michigan, Ann Arbor.

Foster, G. M.
1955 *Contemporary Pottery Techniques in Southern and Central Mexico*. Publication 22. Middle American Research Institute, Tulane University, New Orleans.
1965 The Sociology of Pottery: Questions and Hypotheses Arising from Contemporary Mexican Work. In *Ceramics and Man*, edited by F. R. Matson, pp. 43-61. Viking Fund Publications in Anthropology 41. Wenner-Gren Foundation for Anthropological Research, New York.
1967 Contemporary Pottery and Basketry. In *Handbook of Middle American Indians, vol. 6: Social Anthropology*, edited by M. Nash, pp. 103-124. University of Texas Press, Austin.

Heacock, L. A.
1995 Archaeological Investigations of Three Mesa Verde Anasazi Pit Kilns. *Kiva* 60: 391-410.

Hendry, J. C.
1992 *Atzompa: A Pottery Producing Village of Southern Mexico in the Mid-1950s*. Publications in Anthropology 40. Vanderbilt University, Nashville.

Krotser, P.
1980 Potters in the Land of the Olmec. In *In the Land of the Olmec, vol. 2: The People of the River*, edited by M. D. Coe and R. A. Diehl, pp. 128-138. University of Texas Press, Austin.
1987 Levels of Specialization Among Potters of Teotihuacan. In *Teotihuacan: Nuevos Datos, Nuevas Síntesis, Nuevos Problemas*, edited by E. McClung

de Tapia and E. C. Rattray, pp. 417-427. Universidad Nacional Autónoma de Mexico, Mexico.
Manzanilla, L. (editor)
1993 *Anatomía de un Conjunto Residencial Teotihuacano en Oztoyahualco (Anatomy of a Teotihuacan Residential Compound in Oztoyahualco).* Instituto de Investigaciones Antropológicas, Universidad Nacional Autónoma de Mexico, Mexico.
Marcus, J., and K. V. Flannery.
1996 *Zapotec Civilization: How Urban Society Evolved in Mexico's Oaxaca Valley.* Thames and Hudson, London.
Payne, W. O.
1970 A Potter's Analysis of the Pottery from Lambityeco Tomb 2. *Boletín de Estudios Oaxaqueños* 29: 1-8.
1982 Kilns and Ceramic Technology of Ancient Mesoamerica. In *Archaeological Ceramics*, edited by J. S. Olin and A. D. Franklin, pp. 47-56. Smithsonian Institution Press, Washington, D.C.
1994 The Raw Materials and Pottery-Making Techniques of Early Formative Oaxaca: An Introduction. In *Early Formative Pottery of the Valley of Oaxaca, Mexico*, by K. V. Flannery and J. Marcus, pp. 7-20. Memoir 27. Museum of Anthropology, University of Michigan, Ann Arbor.
Rattray, E. C.
1990 New Findings on the Origins of Thin Orange Ceramics. *Ancient Mesoamerica* 1: 181-195.
Redmond, E. M.
1979 A Terminal Formative Ceramic Workshop in the Tehuacán Valley. In *Prehistoric Social, Political, and Economic Development in the Area of the Tehuacán Valley: Some Results of the Palo Blanco Project*, edited by R. D. Drennan, pp. 111-125. Technical Report 11. Museum of Anthropology, University of Michigan, Ann Arbor.
Rice, P. M.
1987 *Pottery Analysis: A Sourcebook.* University of Chicago Press, Chicago.
1994 The Kilns of Moquegua, Peru: Technology, Excavations, and Functions. *Journal of Field Archaeology* 21: 325-344.
Russell, G. S.
1994 Cerro Mayal, Peru: Moche Ceramic Workshop Excavated. *Backdirt* (Spring): 6-7.
Rhodes, D.
1968 *Kilns: Design, Construction, and Operation.* Chilton Book Company. Philadelphia.
Rye, O. S.
1981 *Pottery Technology: Principles and Reconstruction.* Manuals on Archaeology 4. Taraxacum, Washington, D.C.
Rye, O. S., and C. Evans.
1976 *Traditional Pottery Techniques of Pakistan: Field and Laboratory Studies.* Smithsonian Contributions to Anthropology 21. Smithsonian Institution, Washington, D.C.

Santley, R. S., P. J. Arnold III, and C. A. Pool.
1989 The Ceramics Production System at Matacapan, Veracruz, Mexico. *Journal of Field Archaeology* 16: 107-132.

Sheehy, J. J.
1992 *Ceramic Production in Ancient Teotihuacan, Mexico: A Case Study of Tlajinga 33*. Unpublished Ph.D. dissertation, Pennsylvania State University, University Park.

Shepard, A. O.
1961 *Ceramics for the Archaeologist*. Publication 609. Carnegie Institution of Washington, Washington, D.C.
1963 Beginnings of Ceramic Industrialization: An Example from the Oaxaca Valley. In *Notes from a Ceramic Laboratory 2*, pp. 1-24. Carnegie Institution of Washington, Washington, D.C.

Shimada, I.
1989 Excavation and Experimental Firing of Formative Kilns in Batan Grande, North Coast, Peru, July--September, 1989. Manuscript on file. Peabody Museum, Harvard.

Sinopoli, C. M.
1991 *Approaches to Archaeological Ceramics*. Plenum Press, New York.

Stark, B. L.
1985 Archaeological Identification of Pottery Production Locations: Ethnoarchaeological and Archaeological Data in Mesoamerica. In *Decoding Prehistoric Ceramics*, edited by B. A. Nelson, pp. 158-194. Southern Illinois University Press, Carbondale.
1989 *Patarata Pottery: Classic Period Ceramics of the South-Central Gulf Coast, Veracruz, Mexico*. Anthropological Paper 51. University of Arizona, Tucson.

Stolmaker, C.
1976 Examples of Stability and Change from Santa María Atzompa. In *Markets in Oaxaca*, edited by S. Cook and M. Diskin, pp. 189-207. University of Texas Press, Austin.

Stone, D., and C. Turnbull.
1941 A Sula-Ulúa Pottery Kiln. *American Antiquity* 1:39-47.

Swezey, W. R.
1975 Mound 91, Lambityeco, a Description of an Ancient Kiln. *XIII Mesa Redonda de la Sociedad Mexicana de Antropología*, tomo II: 179-184. Mexico.

van de Velde, P., and H. R. van de Velde.
1939 *The Black Pottery of Coyotopec, Oaxaca, Mexico*. Southwest Museum Papers 13. Southwest Museum, Los Angeles.

Whalen, M. E.
1981 *Excavations at Santo Domingo Tomaltepec: Evolution of a Formative Community in the Valley of Oaxaca, Mexico*. Memoir 12. Museum of Anthropology, University of Michigan, Ann Arbor.

Winter, M. C.
1994 El Proyecto Especial Monte Albán 1992-1994: Antecedentes, Intervenciones y Perspectivas (Monte Albán Special Project 1992-1994: Antecedents, Interventions, and Perspectives). In *Monte Albán: Estudios*

Recientes, edited by M. C. Winter, pp. 1-24. Contribución No. 2 del Proyecto Especial Monte Albán 1992-1994, Oaxaca.
Winter, M. C., and V. Nardin.
1982 Rescate Arqueológico en Loma del Trapiche Guadalupe Hidalgo, Etla, Oaxaca (Salvage Archaeology in Loma del Trapiche Guadalupe Hidalgo, Etla, Oaxaca). *Estudios de Antropología e Historia* 30: 1-14. Centro Regional de Oaxaca, Instituto Nacional de Antropología e Historia, Mexico.
Winter, M. C., and W. O. Payne.
1976 Hornos Para Cerámica Hallados en Monte Albán (Ceramic Kilns Discovered at Monte Albán). *Boletín del Consejo de Arqueología* 16: 37-40. Instituto Nacional de Antropología e Historia, Mexico.

PREHISPANIC KILNS AT MATACAPAN, VERACRUZ, MEXICO

Christopher A. Pool, Department of Anthropology, University of Kentucky, Lexington, KY 40506

ABSTRACT

Investigations in and around Matacapan, Veracruz, Mexico have identified the remains of 40 circular updraft kilns dating to the Classic period. Excavated firing localities provide information on kiln design and construction materials, fuel types, products, and production scale and context. Built with fiber-tempered mud, the prehispanic kilns are similar in design and construction to those used by traditional potters in the surrounding region today. Fragments of kiln material have been recovered from Late Formative contexts, suggesting technological continuity over two millennia. During the Classic Period, kilns were utilized in contexts ranging from small household industries to large nucleated workshops with little variation in design or construction. At Matacapan, control over firing atmosphere appears to have been a more significant factor in kiln use than climate, fuel efficiency, or production intensity.

INTRODUCTION

Archaeologists and engineers (though not necessarily potters) widely regard kilns as a significant improvement over open fires for firing pottery. The many advantages cited for kilns include better control over firing conditions, the possibility of attaining higher temperatures, more efficient use of fuel, and protection from wind and rain (D. Arnold 1985:213; Rice 1987:161; Rye 1981:98). In addition, Philip Arnold (1991:109-113) has argued recently that by using kilns to protect firing loads from drafts, potters in the Sierra de los Tuxtlas of southern Veracruz conserve the amount of space they allot to firing on their small houselots. The variety of advantages attributed to kilns suggests that the origins of kiln technology may have diverse causes in different settings.

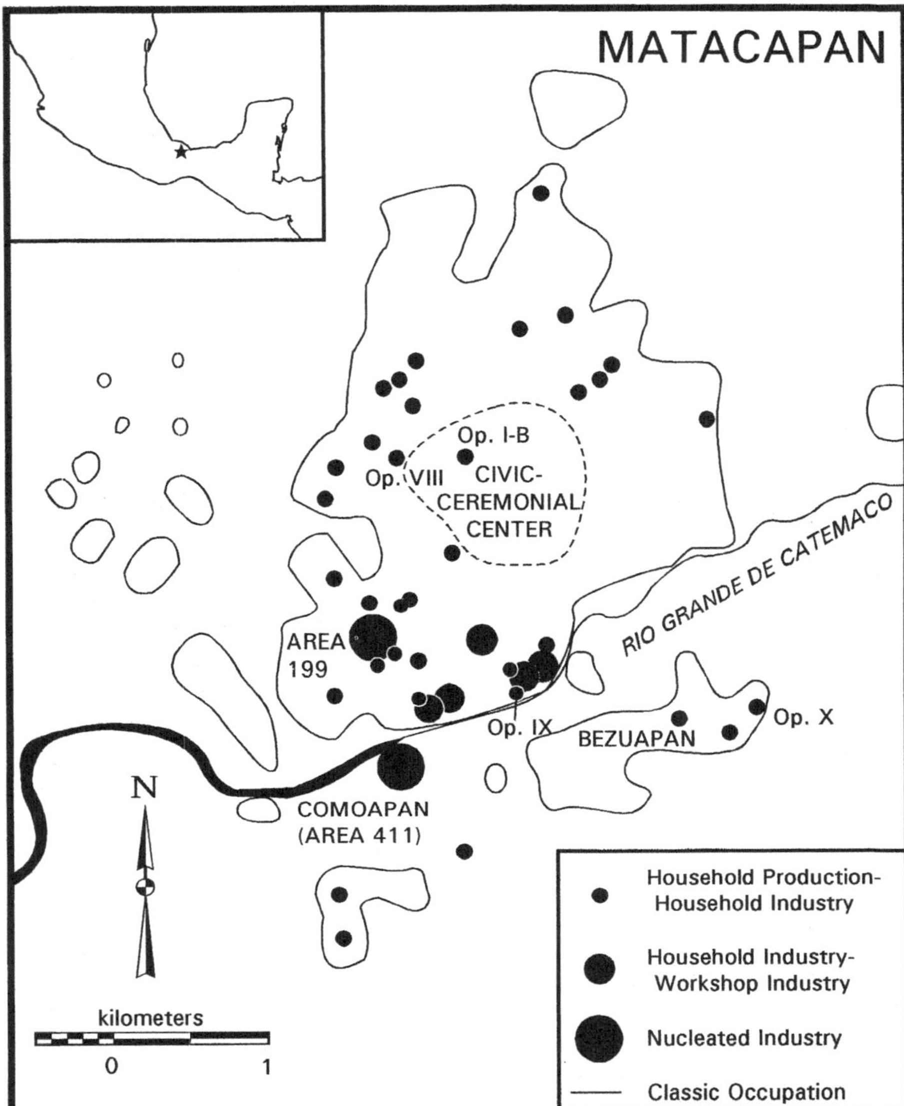

Figure 1. Map of ceramic production areas at Matacapan (redrawn after Santley et al. 1989, with modifications).

Investigations in and around the site of Matacapan, Veracruz, Mexico, provide a case study of the evolution of kiln firing technology in a humid tropical lowland environment (Figure 1). Forty prehispanic kilns have been securely identified at

Matacapan, and the systematic transect survey of the site produced 55 additional surface collections that contained fragments of the fired, fiber-tempered mud used in kiln construction. The ages of the kilns, as determined from their associated ceramics, span the Classic period from A.D. 300 to A.D. 800. Kiln debris recovered from a sealed Late Formative context at the outlying site of Bezuapan suggests that kilns may have been in use by 100 B.C. (Pool 1996). We have found kilns in both domestic and non-domestic contexts, and we interpret kiln contexts as representing a range of production modes that varies from household production to nucleated workshop industries (P. Arnold et al. 1993; Arnold and Santley 1993; Pool 1990:242-248, 322-325, Santley et al. 1989). Previous publications by participants in the Matacapan Project have detailed the overall ceramics production system at Matacapan (Santley et al. 1989), described the nucleated workshop industry in the Comoapan locality (Arnold et al. 1993), identified sources of clay for pottery associated with production areas (Pool and Santley 1992) and discussed the assemblage composition and context of production areas (Pool 1990). In this paper I focus on details of kiln construction and evaluate factors in the origin of kiln technology at Matacapan.

THE CERAMICS PRODUCTION SYSTEM AT MATACAPAN

Forty-two ceramic production areas have been identified at Matacapan. Forty-one of these were identified on the basis of surface collections, and one was discovered in excavation (Santley et al. 1989). Surface indicators of ceramic production consisted of cracked, warped or vitrified ("deformed") waster sherds, kiln construction material ("kiln debris") and high densities of sherds. "High density" was defined as the upper tercile of the distribution of sherd frequencies across the site, or more than 22 sherds per 3 m x 3 m collection square[1] (Pool and Santley 1992:212; cf. Santley et al. 1989:112). These indicators were treated as a polythetic set; a collection had to contain either deformed wasters or both kiln debris and a high sherd density to be counted as a "production square" indicating ceramic manufacture.

Ceramic production areas at Matacapan vary greatly with respect to their context, size, average sherd frequencies, proportion of space devoted to firing activities, and assemblage composition (Santley et al. 1989:115-121). Variation in these parameters suggests differences in scale, intensity, labor organization, and activity segregation from one production entity to another. Monitoring variation among multiple "dimensions of variation" in this manner provides a more precise characterization of ceramic production systems than does pigeonholing them in predefined production modes (Costin 1991:8-18; Pool 1992:277-280). Nevertheless, analysis of variation along these dimensions among the production areas at Matacapan suggests the existence of at least three distinctive production modes, which also correspond to differences in products and reflect the chronological development of the production system. The nomenclature for production modes at Matacapan follows van der Leeuw (1976) and Peacock (1982).

Most of the production areas consist of small groups of 1 to 3 production squares associated with housemounds and domestic refuse or widely dispersed production squares in broad zones of domestic occupation. They exhibit either low average sherd densities or low levels of activity segregation (as measured by the proportion of the area devoted to firing activities), or both (Santley et al 1989:117-118). The characteristics of this group are broadly consistent with small-scale, low intensity domestic production comparable to household production and small household industries in Peacock's (1982) and van der Leeuw's (1976) terminology. These "household production-household industry" areas are widely distributed outside the civic-ceremonial center of the site (Figure 1). Another small production area was identified in excavations (Operation I-B) in a plaza in the civic-ceremonial center of Matacapan. Operation I-B has been interpreted by Santley et al. (1989:108-109, 120-121) as a locus of "tethered" or attached specialist ceramic production for elites, as described in greater detail in the following section.

Five medium to large production areas in the southern part of Matacapan were represented by 4 to 12 production squares and had higher-than-average sherd densities (Figure 1). Although all 5 were associated with domestic occupation, more than 20% of the squares within each production area contained indicators of firing activities. These areas appear to have engaged in ceramic production at a greater scale and intensity than those of the first group, and firing activities appear to have consumed sufficient space to require their segregation from other domestic activities. Depending upon whether labor was organized as a part-time or full-time activity, these areas would represent either large household industries or workshop industries (Santley et al. 1989:118).

The most intensive level of ceramic production is represented by Area 411 in the modern village of Comoapan and Area 199, both of which are located close to permanent water sources and clay outcrops near the southwestern margin of the site (Figure 1). Both contain discrete clusters of kilns and ceramic dumps distributed over 2 ha. in Area 199 and 4 ha in Comoapan. Average sherd densities exceeded 200 sherds per collection and collections from individual waster dumps yielded more than 1000 sherds. Housemounds occurred within Area 199, but none were detected in Comoapan. Recent building activities may have destroyed housemounds in Comoapan, but domestic residues were also scarce, suggesting that production activities were segregated from domestic occupation, though at what distance is uncertain. The size, intensity, and aggregation of production loci in these two areas suggests that they functioned as nucleated workshop industries (Peacock 1982:9). Separate workshops within such nucleated industries may achieve economies of scale by cooperating in resource procurement and distribution of finished products (Pool

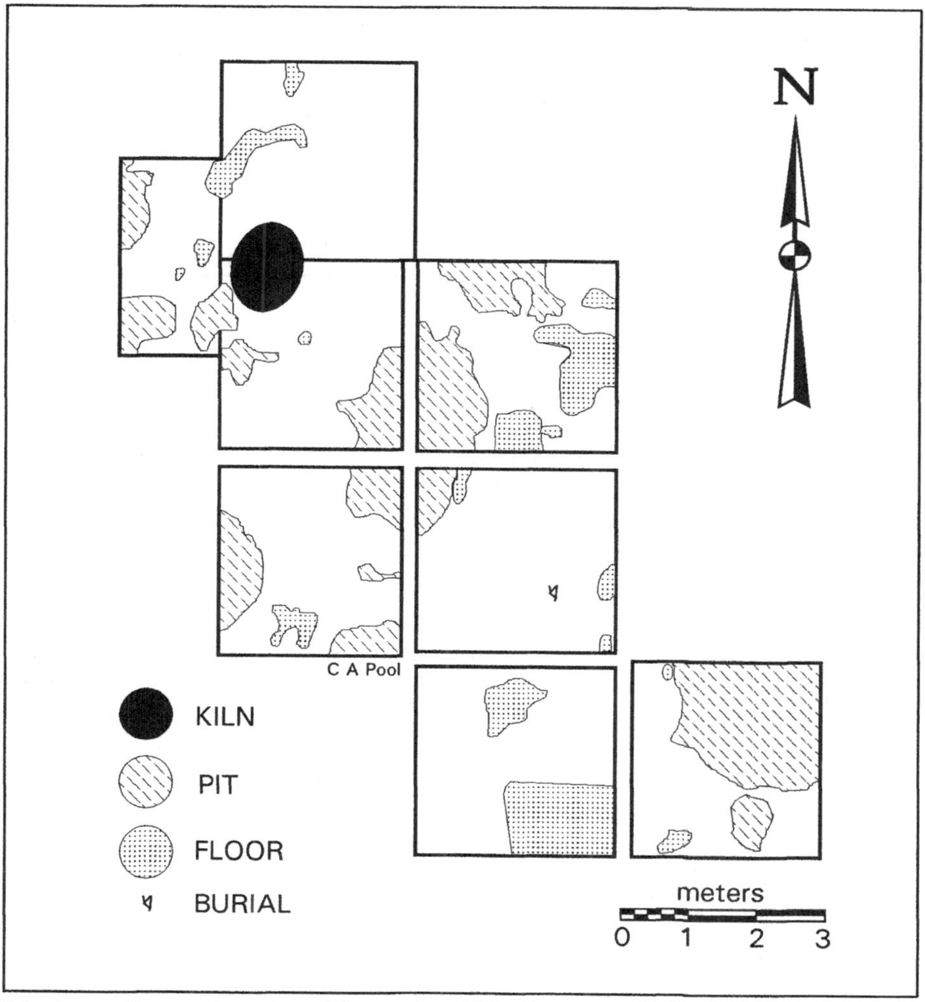

Figure 2. Excavation plan of Operation VIII, Western Locality.

and Santley 1992:214). Alternatively, the Comoapan workshop complex may have functioned as a manufactory under the direction of a general supervisor (Santley et al. 1989:120).

Variation in assemblage composition tends to parallel variation in production mode at Matacapan. Nucleated workshop assemblages are highly skewed towards Coarse Orange jars, while the principal ware in production areas of the large household industry-workshop industry class is Fine Gray. Both of these wares

Table I. Maximum sherd densities in excavated production contexts.

Western Locality (Op. VIII)	838/m^3
Central Locality (Op. I-B)	865/m^3
Southeastern Locality (Op. IX), Lower Unit	2825/m^3
Southeastern Locality (Op. IX), Upper Unit	4650/m^3
Comoapan Locality (Op. VI)	5364/m^3

become increasingly common after A.D. 550, and Fine Gray is most common between A.D. 650 and 800. Household production and small household industry production areas exhibit greater variation in their ceramic assemblages, but Fine Orange service vessels often constitute the principal ware. Coarse Brown ware (mainly jars) is the most common ceramic in six production areas, all of them interpreted as household or small household industry production. As discussed below, a high incidence of fire clouds on Coarse Brown sherds suggests this ware was fired most commonly without the benefit of kilns, and indeed, none of these six areas yielded kiln debris in surface collections. As kiln debris is one criterion we have used to define production areas, its lack of association with Coarse Brown production has undoubtedly biased our sample against domestic manufacture of this common utilitarian ware. Both Fine Orange and Coarse Brown wares were manufactured throughout the Classic Period. In sum, the more intensive modes of production appear to have developed to meet demand for Coarse Orange jars and Fine Gray serving vessels after A.D. 550, but they coexisted with continued household industry production of Fine Orange serving vessels and probably Coarse Brown jars as well (Pool 1990:332-333).

CONTEXTS OF EXCAVATED KILNS

Data on kiln construction at Matacapan derives from excavations in four localities (Figure 1). The Western Locality (Operation VIII) contained the destroyed remains of a single kiln in the patio of a nonelite residence (Figure 2). Three large trash pits and six smaller pits contained a mixture of household refuse and ceramic production residues, including waster sherds. The largest pit was more than 2.25 m in diameter and 2 m deep. Although these pits contained the highest ceramic densities in the Western Locality, the maximum sherd density of 838/m^3 was the lowest of all the excavated production localities (Table I). The Western Locality probably represents the remains of small-scale production for the household and limited

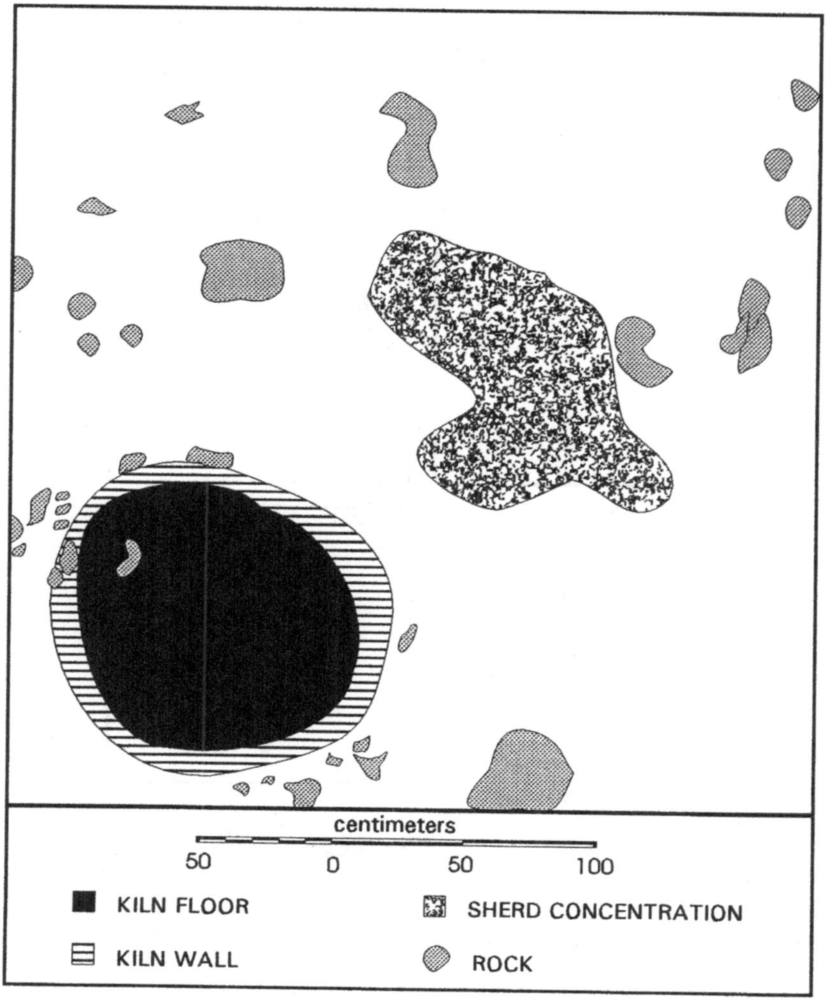

Figure 3. Excavation plan of Operation I-B, Central Locality.

exchange, that is, household or household industry production. Production appears to have focused primarily on Fine Orange serving vessels, and the associated ceramics suggest a date in the Early Classic (A.D. 300-450).

A 3 x 3 m excavation unit in the Central Locality (Operation I-B) uncovered the firebox of a second kiln within a plaza surrounded by pyramids and elite residential mounds (Figure 3). The sherd density of 865/m^3 around the kiln only slightly exceeded the maximum sherd density in the Western Locality, but the limited

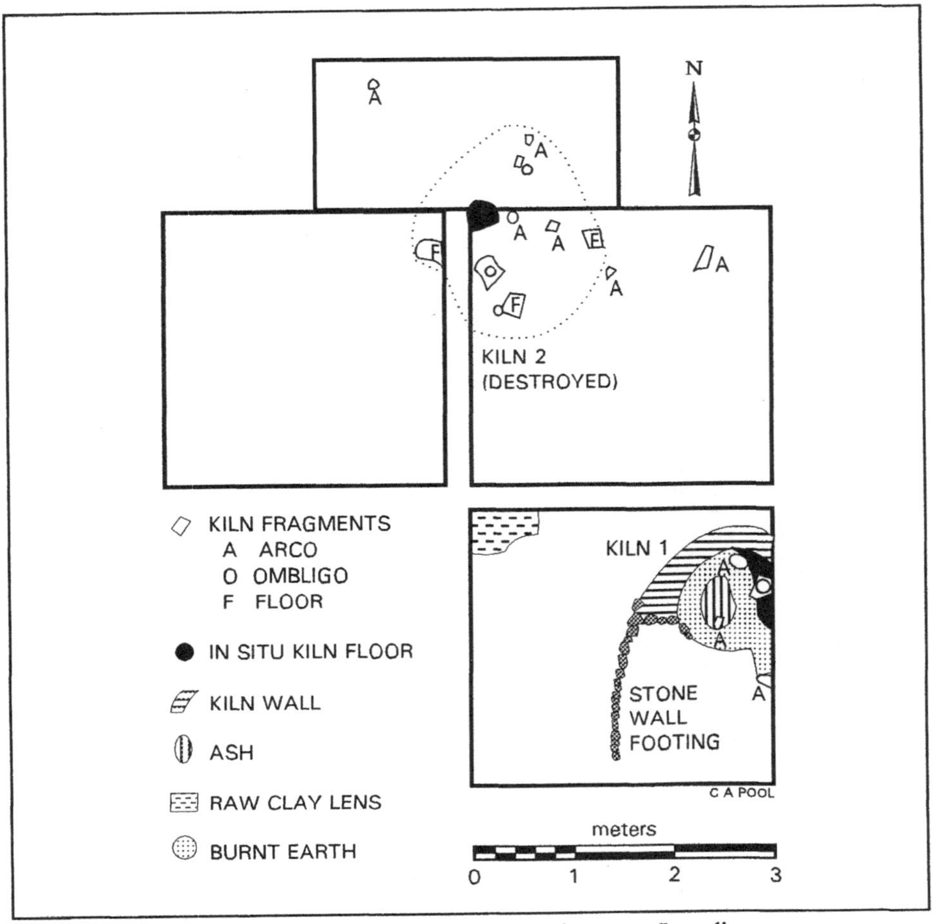

Figure 4. Excavation plan of Operation IX, Southeastern Locality.

exposure produced no household refuse. Waster sherds and pieces of vitrified clay were common, and several small burnishing stones were recovered. Production focused on neckless Coarse Orange jars (30.3% of rim sherds), burnished Brown-slipped Fine Orange serving vessels (21.7%) and Coarse Brown jars (12.0%). Both the Coarse Orange and Brown-slipped Fine Orange types are most common at Matacapan in the late Middle Classic period (A.D. 550-650), but Brown-slipped Fine Orange type comprises less than 2% of the total late Middle Classic assemblage at the site (Ortiz and Santley 1989:Figure 2.8). The context of the central locality, its unimpressive sherd density, and the specialized nature of its assemblage suggest that

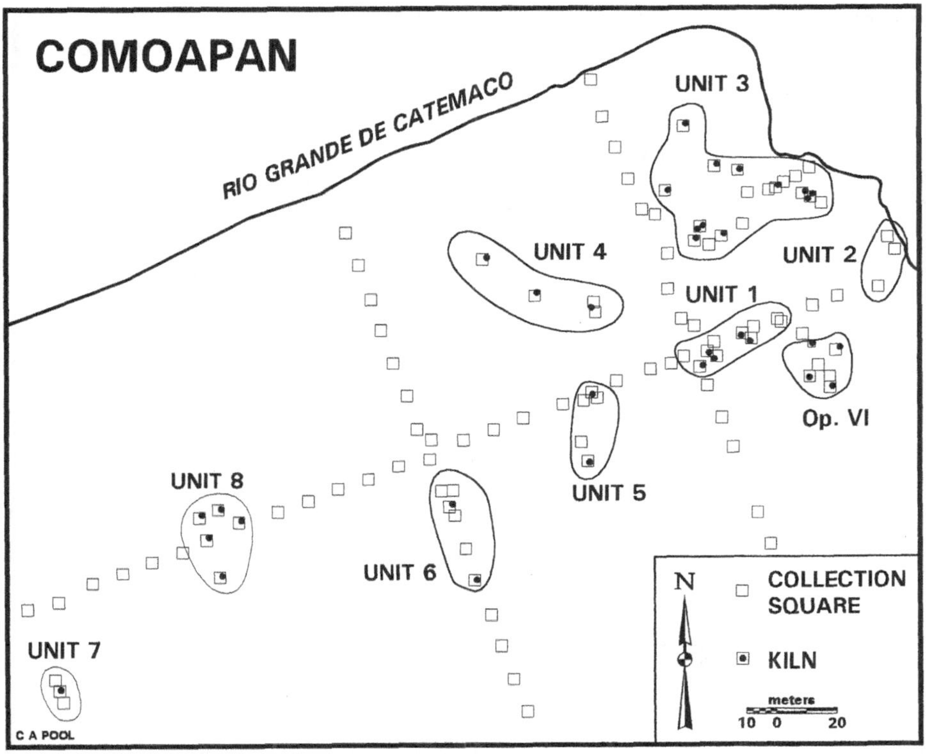

Figure 5. Surface collections, kilns and production units at the Comoapan production locality (after Arnold et al. 1993:Figure 3 and author's field notes).

it served as the workshop of an attached specialist who produced pottery at a relatively small scale for an elite client.

Excavations in the Southeastern Locality (Operation IX) encountered the remains of two kilns associated with a stratigraphically superimposed series of waster dumps to the east of a nonelite residential mound (Figure 4). The lower waster dumps contained approximately equal amounts of Fine Gray (40.3%) and Fine Orange rim sherds (37.2%). The upper dump contained substantially more Fine Gray rim sherds (61.5%) and fewer Fine Orange rim sherds (11.5%). Sherd densities in the upper dump were higher (at 4650/m^3) than in the lower dump (2825/m^3). The changes in the prevalence of wares and the densities of sherds may reflect the development of a late Middle Classic household industry into a Late Classic workshop specializing in Fine Gray production.

The largest and most specialized of the Matacapan production localities covers about 4 ha within the modern village of Comoapan. Survey discovered 36 kilns

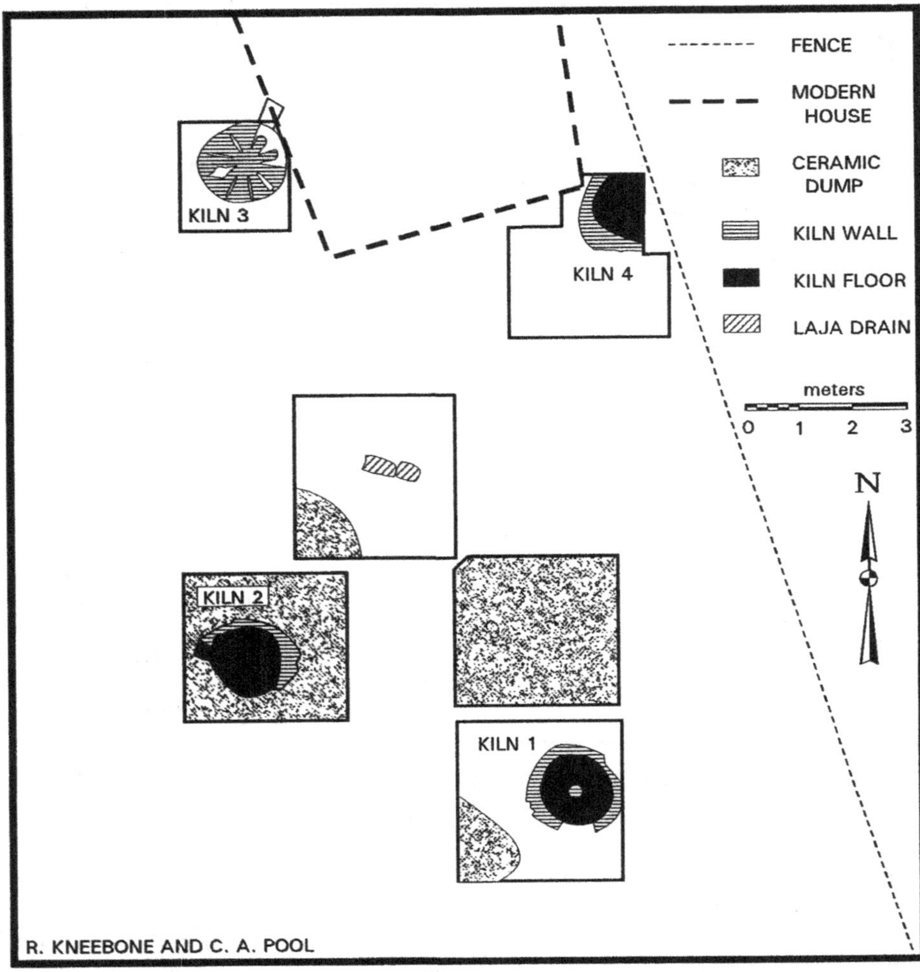

Figure 6. Excavation Plan of Operation VI in the Comoapan Locality (after Arnold et al. 1993:Figure 4 and author's field notes).

associated with waster dumps exposed in the streets and patios of the village (Figure 5). Four of these kilns were excavated in Operation VI (Figure 6). Ceramic densities in the waster dump associated with the excavated kilns attained 5364/cu m. Coarse Orange jars with and without necks comprised 63.4% of all excavated rims in Comoapan. Domestic refuse and residential architectural features were absent from the Comoapan production complex, suggesting that potters lived at some remove from their workshops. The spatial extent, high ceramic densities, high level of

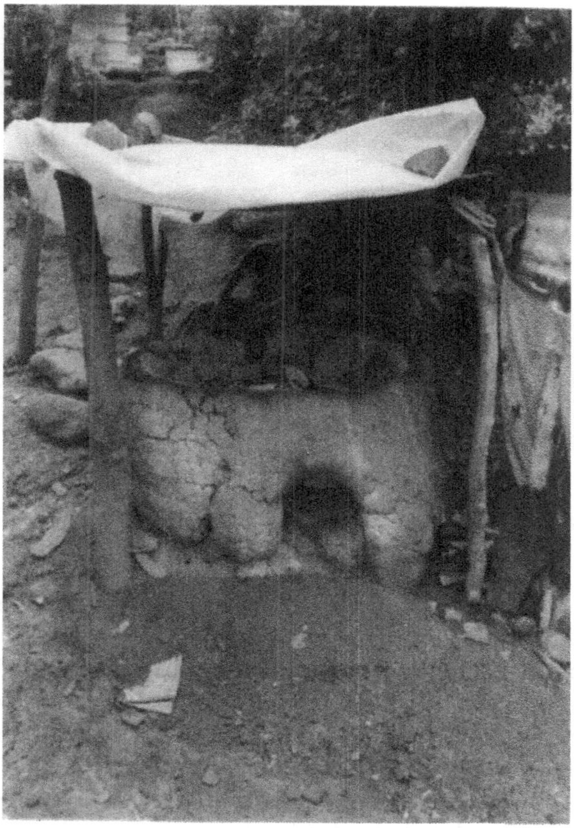

Figure 7. Modern kiln, San Isidro Texcaltitan, Municipio of San Andres Tuxtla, Veracruz, Mexico. Photo by author.

product specialization and lack of evidence for household occupation within the Comoapan locality suggest large scale specialized production in a nucleated workshop industry (Arnold et al. 1993; Pool 1990:247).

KILN CONSTRUCTION

At Matacapan, the identification of features as kilns relies on evidence for intensive heating, association with production residues and failed kiln loads, and similarity in materials and form with kilns used by contemporary potters in the Sierra de los Tuxtlas (P. Arnold 1991:54-56, 145). Although most potters in the Tuxtlas use open fires, the potters of San Isidro Texcaltitan, in the *municipio* of San Andrés Tuxtla, fire pottery in small, cylindrical, double-chambered updraft kilns (Figure 7).

Figure 8 Prehispanic Kiln 3 in Operation VI, Comopan Locality. Photograph by Philip J. Arnold III (Arnold et al. 1993: Figure 6).

The walls of the kilns are built of alternating layers of fiber-tempered mud (adobe) and small stones raised on a foundation of irregular stones placed directly on the ground surface. A removable covering of plastic or tar paper supported by four poles protects the kiln from rain. Access to the firebox is through a small opening topped by a rough stone lintel. A stone called the *ombligo* set in the center of the firebox supports a grate of scrap metal and large sherds. As described by P. Arnold (1991:54), 30 to 40 vessels are loaded into the firing chamber through the open top of the kiln. Large and enclosed vessels (*cazuelas* and *ollas*) are loaded first, smaller *tecualones* and *maceteras* form the second level, and flatter *comales* are placed on top to help retain heat. The top of the kiln is then closed off with large sherds, but no mud or clay is used to cap the load. The high temperatures achieved in the firebox create a vitrified interior surface on the lower walls of the kiln, but the upper walls and exterior surfaces are less highly fired. P. Arnold (personal communication 1991) reports that the kilns last up to three years before they must be rebuilt.

The best-preserved of the prehispanic Matacapan kilns follows a similar design. Kiln 3 in Operation VI in the Comoapan locality was built with puddled fiber-tempered adobe walls set directly on the ancient ground surface (Figure 8). An adobe post in the form of an inverted truncated cone was set in the center of the kiln. This *ombligo* supported a grate formed by 9 trapezoidal plano-convex adobe *arcos* arranged radially about the *ombligo* and cemented to it with daubed adobe (Figure 9). The outer ends of the *arcos* were set on the firebox wall. On the northeast side of the kiln, a tunnel-like entrance 30 cm high, 30 cm wide, and 50 cm long provided access to the fire chamber. A large basalt stone blocked the entrance to this stoking tunnel. All surfaces within the firebox were highly vitrified, to the point that drops of slag formed on the underside of the arcos and ran down the firebox walls. The walls of the upper chamber were no longer intact, but the unsmoothed exterior ends of the arcos may indicate the area of their articulation with the firebox.

The interior walls of the firebox sloped outward from a diameter of 113 cm at the floor to 149 cm at the base of the arcos, 35 cm above the firebox floor. The original exterior dimensions of the kiln are somewhat less certain, as is the thickness of the walls, because the exterior surfaces of the walls were less highly fired. The remaining fired portion of the firebox walls were between 11 and 24 cm thick, and the exterior diameter varied from 145 to 172 cm.

Table II presents data on the dimensions and construction techniques of all excavated kilns at Matacapan. These data indicate that Comoapan Kiln 3 was fairly typical, but that construction techniques and dimensions varied even within the excavated Comoapan workshop. Kiln 1 was also built directly on the ground surface, but stones were incorporated into the foundation of the adobe walls. In contrast, the fireboxes of Kilns 2 and 4 were partially excavated into the sloping ground surface, perhaps primarily to provide a level floor. Firebox mouths in the excavated Comoapan workshop were oriented to the northeast, west, and southeast, not with regard to the prevailing northerly winds, but on the downslope sides of the kilns, possibly to prevent flooding of the firebox due to rainwater runoff. Although temporary coverings may have protected the kilns from falling rain, no postmolds or other evidence for such structures were encountered in the Comoapan excavations.

Elsewhere in Matacapan, the badly destroyed kiln in the Western locality was built directly on a natural surface of consolidated volcanic ash (*laja*). Pieces of this material and stones were incorporated into the adobe walls, only a small fragment of which was intact.

In the Southeastern Locality, Kiln 1 was partly excavated into a steep slope. The collapsed firing chamber walls and front wall of the firebox were made of adobe, but again, stones appear to have been incorporated in the construction of the kiln. In addition, small stones appear to have been used for the footing of the perishable walls of a small structure in front of the kiln. The adobe ombligo and arcos of the kiln were found collapsed in the firebox. Kiln 2, which lay higher in the excavation, was badly destroyed by plowing but appears to have been built directly on a former ground surface. The adobe ombligo and arcos were scattered around a small intact portion of the vitrified firebox floor.

The kiln in the central locality is unusual in that the firebox consisted of a shallow pit. Stones and fragments of fired adobe on the margins of this pit represent the base of the firing chamber walls.

Table II. Kiln Dimensions[1] and Construction

Locality	Interior Diameter	Exterior Diameter	Firebox Depth	Firebox Volume	Wall Thickness	Construction
Bezuapan (Op. X)	--	--	--	--	--	adobe wall fragments
Western (Op. VIII)	120	--	--	--	--	surface kiln, adobe, laja and stones
Central (Op. I-B)	109	126	40	0.37	12	semisubterranean, adobe and stones
Comoapan (Op. VI)						
Kiln 1	138	171	27	0.40	21	surface kiln, adobe and stones
Kiln 2	125	--	44	0.54	22	partly dug into slope, adobe
Kiln 3	131	158	35	0.47	18	surface kiln, adobe
Kiln 4	>122	--	30	--	21	partly dug into slope, adobe
Southeastern (Op. IX)						
Kiln 1	109	--	46	0.43	20	partly dug into slope, adobe and stones
Kiln 2						destroyed, adobe ombligo, arcos, wall
San Isidro (Modern)[2]	85	104	27	0.15	--	surface kilns, adobe, stones, and scrap iron

[1]Dimensions are averages in cm of maximum and minimum measurements, except firebox volumes, which are in m^3.
[2]Averages from P. Arnold 1991:54.

Kiln dimensions are remarkably consistent across time and production context (Table II). Average interior firebox diameters range from 109 to 138 cm, and firebox depths range from 27 to 46 cm. These figures also provide estimates of total firebox volume (including the central ombligo) of 0.37 cu m to 0.54 cu m. The attached specialist workshop in the Central Locality and the unattached workshop of the Southeastern Locality contained the smallest kilns, and the nucleated industry workshop in Comoapan contained the largest kiln, but overall, the excavated kilns do not exhibit consistent variations with production mode, principal products, or chronological period.

The distinctive material employed in kiln construction serves as a good indicator of kilns in surface contexts. As opposed to daub used in house construction, the adobe used in kilns at Matacapan invariably was tempered with vegetable fibers, lacks impressions of poles or saplings, and exhibits either vitrification on one surface or differential firing in cross-section as a result of repeated exposure to high heat on the interior surface. On occasion, specific parts of the kiln, such as the ombligo or the arcos, can be identified in surface collections. Where the ground surface has not been disturbed by plowing, as in the streets and yards of Comoapan, the walls of the kilns' firebox appear as a circle of fired adobe. Evidence of the usefulness of these criteria for identifying kilns is provided by the discovery of *in situ* kilns in the Western and Southeastern localities, where the location of Operations VIII and IX were chosen because of the presence of kiln debris in surface collections.

FUEL

The best information on the fuel employed in the Matacapan kilns comes from Kiln 1 in the Southeastern Locality. The firebox of this kiln contained a light fibrous ash similar to that produced by burning palm leaf stalks, a common fuel for firing pottery in the Tuxtlas today (P. Arnold 1991:25-26). Carbonized seeds of the *coyol redondo* palm (*Acrocomia mexicana* or *Scheelea liebmanii*) recovered from the firebox and the area in front of this kiln confirm the burning of palm, although it is not clear whether the seeds were used intentionally for fuel or included incidentally. In the fireboxes of other kilns at Matacapan, charcoal is scarce, and a similar fibrous ash is typically present, suggesting that palm may have been the generally preferred fuel.

The use of palm is somewhat surprising because its open structure and low specific gravity make palm a notoriously poor fuel. Palm burns rapidly and has a low heating value as compared to hardwoods such as those of the genera *Albizia*, *Guarea*, *Malmea*, *Pterocarpus*, and *Trichilia*, which are found in the high evergreen *selvas* (forests) of the Sierra de los Tuxtlas, and which are widely used for fuel in Central America (Gómez-Pompa 1973:111; Jones and Pérez 1981; Jones and Otavalo 1981). Previously I have argued that the use of palm in kilns at Matacapan resulted from competition between pottery production and cooking for hardwood fuel (Pool 1991). Today in the similar environment of the Nicoya Peninsula, the consumption of fuel averages 2.1 kg per person per day on small *fincas*, where it is used primarily for cooking (Jones and Otavalo 1981). Based on conservative estimates of population and generous estimates of pottery demand, I calculate that total domestic fuel consumption at Matacapan over the course of the Middle Classic would have exceeded 365,230,000 kg, while pottery production for the same population would have consumed a maximum of 7,750,000 kg of fuel, or about 2.1% of domestic fuel consumption.

As the growing population of Matacapan placed increasing pressure on the surrounding evergreen selvas for agricultural land and cooking fuel, potters may have opted for palm as a less efficient but adequate fuel, near at hand, that was not suitable for cooking, as many contemporary potters in the Tuxtlas do today. Today two palm species are widely called "coyol" in Mexico, *Scheelea liebmanii* and *Acrocomia mexicana*. Both produce "branches" that die and can be removed without harming the tree. Moreover, both produce edible seeds, their fronds provide thatch for roofing material and their spathes are sometimes used as containers (Coe and Diehl 1980:92). Describing uses of *Scheelea liebmanii*, Coe and Diehl (1980:92) report that inhabitants of southern Veracruz eat the young sprouts, the heart, and the seeds, drink the juice of the trunk, prepare cooking oil from the seeds, and make a soap by burning the seeds and mixing the ashes with lard. Protected or cultivated stands of these palms in Matacapan would therefore have provided several economic advantages, only one of which would have been fuel for pottery production.

DISCUSSION: WHY KILNS AT MATACAPAN?

Returning to a theme expressed in the opening line of this paper, although archaeologists and engineers extol the virtues of kilns, their advantages are by no means universally appreciated by traditional potters, many of whom make not only serviceable utilitarian pottery, but also aesthetically extraordinary works of art using open and pit firing techniques. A case in point is the Sierra de los Tuxtlas, where today potters in different communities use kilns or open fires to make the same range of forms (P. Arnold 1991). Furthermore, the commonly cited advantages of kilns are most evident in kilns that are more complex than the simple circular updraft kilns used by prehispanic and contemporary potters in the Tuxtlas. Given the additional labor required, it is reasonable to ask why potters developed kilns in the first place, particularly in the New World, where until recently prehispanic kilns were thought not to exist (e.g., D. Arnold 185:218).

The preceding discussion of pressures on fuel sources suggests the possibility that a need for increased fuel efficiency may have provided the impetus for kiln technology at Matacapan. This argument fails for two reasons. First, the earliest well-documented kiln at Matacapan dates to the Early Classic, when Matacapan was small, with an estimated population of fewer than 1250 people in about 43 ha, and regional population densities were on the order of 7 people per sq km (Pool 1991; Santley 1991:7). Population densities were higher in the Late Formative period, which yielded possible kiln wall fragments at the village site of Bezuapan, but the maximum estimated regional density of 13.9 per sq km (Santley 1991:5) seems insufficient to stress fuel supplies.

Second, many authors have noted that kilns in general and simple updraft kilns in particular may be less fuel efficient than open firing because heat is expended in heating the kiln structure and much heat is lost through the top of the kiln (Rice 1987:160; Rye 1981:25). In the Tuxtlas, P. Arnold (1991b:17) found that open firings produced an average of about 51 usable vessels per *carga* of fuel (about 35 kg), while kiln firings produced only about 32 usable vessels per *carga*, making the open firings about 88% more fuel efficient in terms of their vessel-to-fuel ratio. It therefore appears that if fuel efficiency were the overriding concern at Matacapan, bonfires would have been favored over kilns.

Neither do the higher temperatures achievable in kilns appear to have provided the impetus for their use at Matacapan. Clay sources utilized by the Matacapan potters have been identified by X-ray fluorescence spectrometry (Pool and Santley 1992). Experiments on samples of these clays indicate that firing in an oxidizing atmosphere produces colors similar to those of archaeological Fine Orange ceramics from Matacapan at between 750°C and 800°C. Higher temperatures were achieved at points in the prehispanic kilns, but they often produced warped and vitrified sherds. Simple updraft kilns do not appear to provide a clear advantage over open firings in achieving this temperature range. Using flame color as an indicator, P. Arnold (1991a:55) estimated maximum temperatures in the modern Tuxtlas updraft kilns to be a little above 800°C and those in open firings to be between 700°C and 850°C. According to Gosselain (1992:244), reported temperature ranges for open, pit, and updraft kiln firings overlap considerably, with all types achieving temperatures of 600°C to 900°C. Furthermore, temperatures in updraft kilns commonly vary by as much as 100°C to 300°C even within the same firing (Nicholson and Patterson 1989:79-83, Rye 1981:Table 3; Sheehy 1988:Table 2).

D. Arnold (1985:215-218) argues that the cross-cultural distribution of kilns reflects their effectiveness in mitigating the effects of weather and climate. Certainly the Sierra de los Tuxtlas, where potting communities experience annual rainfall totals of 2000-2500 mm, would appear to conform to this pattern. Although the open tops of the Tuxtlas kilns provide little protection from rain without a protective structure, the downslope orientation of firebox doors at Matacapan would have provided some protection from runoff on the ground. A climatic explanation, however, is weakened by the observation that modern Tuxtlas potters typically use open firings in areas that have only marginally less rainfall than San Isidro Texcaltitan, where kilns are used (P. Arnold 1991:58-59). Furthermore, the only evidence for a structure directly associated with a kiln at Matacapan is the wall footing in the Southeastern Locality, which appears to provide more protection for the potter than the kiln.

P. Arnold (1991:107-109) advances the argument that kilns, by mitigating microenvironmental variations, such as shifts in the wind, allow potters on small houselots to conserve space, and this does appear to be one advantage for the potters of San Isidro Texcaltitan, whose houselots are smaller, on the average, than those in other pottery-making communities. No such constraints on houselot size appear to have existed in Early Classic Matacapan or Late Formative Bezuapan, however (Pool 1991, 1996; Santley 1991:7).

Finally, kilns are often associated with more intensive production, particularly in the minds of archaeologists (e.g., Peacock 1982:8-10, Sinopoli 1991:101, Santley et al. 1989, van der Leeuw 1984). Cross-culturally, this association has considerable validity, but it does not hold for prehispanic Matacapan, where kilns are found in both low intensity and high intensity production contexts, nor for the modern Tuxtlas, where potters using kilns and open fires produce pottery at similar rates (P. Arnold 1991:27-28)..

The question then remains, why did the ancient potters of Matacapan adopt kiln technology? The principal advantage of the kind of simple updraft kilns used at Matacapan appears to be control over firing conditions, specifically control over the firing atmosphere and the *rate* of heating. In comparing thermometric data from 27 ethnographic firings reported from various parts of the world, Gosselain (1992:246) found that, on the average, open and pit firings required only 22 min and 41 min, respectively, to achieve maximum temperature. These figures increased to 60 min and 114 min when sherds were used to separate the vessels from the fuel. In marked contrast, updraft kilns required an average of 259 min to achieve maximum temperature. Gosselain (1992:244-246) concludes that "the heating rate is by far the most specific parameter for each of the firing techniques" that he considered, and that this observation "might explain, more than temperatures reached or thermic homogeneity, the choices and innovation in the field of firing techniques."

In the summer of 1996, a potter of San Isidro Texcaltitan echoed Gosselain's hypothesis when I asked her why she and others in San Isidro used kilns instead of firing pottery in the open as people do in Sehualaca. First she expressed surprise that other people did not use kilns. After a few moments' thought (and with no further prompting from me) she explained that in San Isidro they used firewood *(leña) and heated the kilns slowly ("con calma")* for about 1 1/2 hours, after which they add firewood until the upper portion glows red. She added that she did not believe she could heat the pottery as slowly in an open fire, and that the vessels would be likely to break. I then asked why she did not use palm fronds instead of firewood. She responded that she did, occasionally, but that the palms were on private land. Although this potter included the type of fuel in her response, she emphasized the slow rate of heating and she appears to consider firewood as a preferable fuel for controlling the heating schedule, while indicating that palm fronds could be used and occasionally were.

Citing Gibson and Woods (1989), Gosselain (1992:257) also notes that only coarse pottery can be fired under the rapid heating conditions of open firing, because the open texture of coarse pottery allows the steam to escape, whereas steam may build up in fine-paste pottery and cause it to shatter. At Matacapan, most of the Classic period pottery belongs to four wares: Fine Orange, Fine Gray, Coarse Orange, and Coarse Brown. Fire clouds appear on the surfaces of 22.9% of a random sample of 367 Coarse Brown sherds from excavated production contexts, but are much less common on the other three wares (7.6% of 877 Fine Gray sherds, 5.2% of 991 Fine Orange sherds, and 1.2% of 598 Coarse Orange sherds) (Pool 1990:Table 14). These data indicate that less control was exercised over the atmosphere in which Coarse Brown pottery was fired, suggesting that it was more often fired in the open, and the other wares were usually kiln-fired. Moreover, surface collections in production areas that emphasized Coarse Brown manufacture are notable for their lack of kiln debris. Coarse Brown is by far the coarsest of the Classic and Late Formative period wares, containing more temper (28.1%-32.7% by volume) and coarser temper (modal maximum aplastic size = coarse sand) than the other wares (Pool 1990: Tables 11 and 13). In contrast, the maximum aplastic size in most Coarse Orange sherds is medium sand, and the average aplastic density ranges from 11.3% to 26.1% by volume. The fine paste ceramics are untempered, but have minor amounts of natural calcareous and other aplastics.

Untempered fine paste pottery first appears in the central Tuxtlas in the Late Formative period and becomes common in the Early Classic. Similarly, the first direct evidence for kiln firing occurs in the form of kiln wall fragments from a Late Formative context at the site of Bezuapan, which was sealed by the ash from a volcanic eruption. The earliest *in situ* kiln dates from the Early Classic at Matacapan, after which kilns become common. The development of kiln technology in the central Sierra de los Tuxtlas therefore appears to coincide with the appearance and increasing popularity of ceramics whose fine paste and uniform surface hues required the greater control over heating rate and firing atmosphere afforded by kilns. That control of the firing atmosphere was a concern for the ancient kiln-using potters is further suggested by the stone that was found blocking the entrance to the firebox in Kiln 3 at Comoapan.

CONCLUSION

In this paper I have described the construction of the prehispanic kilns at Matacapan, as well as the variation in the products of those kilns and the organization of ceramic production. The Matacapan kilns were simple, rather small, circular updraft kilns made primarily of fiber tempered mud, or adobe, and built either entirely above ground, or partially excavated into the usually sloping ground surface. Kilns were used principally for firing fine paste wares in both oxidizing and reducing atmospheres, as well as oxidized jars with volcanic ash temper. In contrast, coarser brown jars tempered with sand, which often included quartz and feldspar clasts in addition to volcanic ash, appear to have usually been fired in the open, resulting in a higher incidence of fire clouds due to less control over firing atmosphere. Kilns of the same design were used in the entire range of production contexts at Matacapan, from small-scale domestic production loci to large-scale nucleated industries. They were definitely in use by the Early Classic period (A.D. 300-450), and perhaps as early as the Late Formative period (400 B.C.-A.D. 100), coinciding with the appearance and expansion of the fine paste ceramic tradition at the site. Kiln technology continued in conjunction with open firing through the Classic period, and both techniques continue to be used by modern potters in the region. Given these data, and considering both the advantages and disadvantages of simple updraft kilns, it appears that at Matacapan the innovation of kiln firing was a response to the need for greater control over firing atmosphere, and particularly the rate of heat increase, required in the production of fine paste ceramics.

The evolution of ceramic pyrotechnology reflects solutions to specific problems that vary from culture to culture depending on the raw materials available, the uses to which the products are put, and the organization of pottery production. The Matacapan case may or may not provide a model for the invention and adoption of kilns in other cultural contexts. The important point is that now that the idea that kilns were introduced to the New World by Europeans has been put to rest, Mesoamerica provides fertile ground for research into the independent invention of kiln technology.

ACKNOWLEDGMENTS

The field research reported in this paper was supported by the National Science Foundation through grants to the University of New Mexico (Robert Santley, principal investigator) and Tulane University. Permits were granted by the Instituto Nacional de Antropología e Historia of Mexico. I thank Izumi Shimada, Prudence Rice, Reinhard Bernbeck, and Michael Malpass for their helpful comments on a draft of this paper.

NOTES

[1]The surface collection program implemented at Matacapan was multi-staged. The first stage employed a systematic interval transect strategy in which 3 m x 3 m collection squares were placed at 13 m intervals along transects spaced 50 to 100 m apart (Santley et al. 1984:10-14). Four ceramic production areas were later resampled using a more intensive stratified random strategy (Pool 1990:196-204). In addition, each kiln and waster dump in the large Comoapan workshop complex was sampled with a provenienced collection square (Arnold et al. 1993:177). The surface data discussed in this article derive from the systematic transect sample, unless otherwise stated.

REFERENCES

Arnold, D.

1985 *Ceramic Theory and Cultural Process.* Cambridge University Press, Cambridge.

Arnold, P.

1991a *Domestic Ceramic Production and Spatial Organization: A Mexican Case Study in Ethnoarchaeology.* Cambridge University Press, Cambridge.

Arnold, P.

1991b Pottery Firing and Ceramic Ecology in the Tuxtlas Mountains, Veracruz, Mexico. Manuscript in possession of the author.

Arnold, P. J. III and R. S. Santley

1993 Household Ceramics Production at Middle Classic Period Matacapan. In *Prehispanic Domestic Units in Western Mesoamerica,* edited by R. S. Santley and K. G. Hirth, pp. 227-248. CRC Press, Boca Raton.

Arnold, P. J. III, C. A. Pool, R. R. Kneebone, and R. S. Santley

1993 Intensive Ceramic Production and Classic-Period Political Economy in the Sierra de los Tuxtlas, Veracruz, Mexico. *Ancient Mesoamerica* 4:175-191.

Coe, M. D. and R. A. Diehl

1980 *In the Land of the Olmec, Volume 2: The People of the River.* University of Texas Press, Austin.

Gibson, A. M. and A. J. Woods

1989 *Pottery for the Archaeologist.* Columbia University Press, Columbia.

Gómez-Pompa, A.

1973 Ecology of the Vegetation of Veracruz. In *Vegetation and Vegetational History of Northern Latin America,* edited by A. Graham, pp. 73-148. Elsevier, Amsterdam.

Gosselain, O. P.
1992 Bonfire of the Enquiries. Pottery Firing Temperatures in Archaeology: What For? *Journal of Archaeological Science* 19:243-259.

Jones, J. R. and A. Otavalo
1981 Diagnóstico socio-económico sobre el consumo y producción de la leña en fincas pequeños en Nicaragua. IRENA-CATIE_ROCAP, Turrialba, Costa Rica.

Jones, J. R. and Pérez
1981 Diagnóstico socio-económico sobre el consumo y producción de leña en Honduras. CATIE-ROCAP, Turrialba, Costa Rica.

Nicholson, P. T. and H. L. Patterson
1989 Ceramic Technology in Upper Egypt: A Study of Pottery Firing. *World Archaeology* 21(1):71-86.

Ortiz Ceballos, Ponciano and Robert Santley
1989 La Cerámica de Matacapan. Ms. on file, Department of Anthropology, University of New Mexico, Albuquerque, New Mexico.

Pool, C. A.
1990 Ceramic Production, Resource Procurement, and Exchange at Matacapan, Veracruz, Mexico. Ph.D. dissertation, Department of Anthropology, Tulane University. University Microfilms International, Ann Arbor.
1991 Quest for Fire: Fuel Costs and Pottery Manufacture at Matacapan, Veracruz. Paper presented at the 23rd annual Chac Mool Conference, Calgary, Alberta.
1996 The Spatial Structure of Formative Houselots at Bezuapan, Veracruz, Mexico. In *Olmec to Aztec: New Research on the Ancient Gulf Coast,* edited by Barbara S. Stark and Philip J. Arnold III. University of Arizona Press, Tucson. In press.

Pool, C. A. and R. S. Santley
1992 Middle Classic Pottery Economics in the Tuxtla Mountains, Southern Veracruz, Mexico. In *Ceramic Production and Distribution: An Integrated Approach,* edited by G. J. Bey III and C. A. Pool, pp. 205-234. Westview Press, Boulder, Colorado.

Peacock, D. P. S.
1982 *Pottery in the Roman World.* Longman, London.

Redmond, E. M.
1979 A Terminal Formative Ceramic Workshop in the Tehuacan Valley. In *Prehistoric Social, Political, and Economic Development in the Area of the Tehuacan Valley: Some Results of the Palo Blanco Project,* edited by R. Drennan, pp. 111-215. Museum of Anthropology, University of Michigan, Ann Arbor.

Rice, P. R.
1987 *Pottery Analysis: A Sourcebook.* University of Chicago Press, Chicago.
Rye, O. S.
1981 *Pottery Technology.* Taraxacum Inc., Washington, D.C.
Santley, R. S.
1991 Final Field Report: Tuxtlas Region Archaeological Survey, 1991 Field Season. A Report to the National Science Foundation. Manuscript on file, Department of Anthropology, University of New Mexico, Albuquerque, New Mexico.
Santley, R. S., P. J. Arnold III and C. A. Pool
1989 The Ceramic Production System at Matacapan, Veracruz, Mexico. *Journal of Field Archaeology* 16:107-132.
Sheehy, J. J.
1988 Ceramic Ecology and the Clay/Fuel Ratio: Modeling Fuel Consumption in Tlajinga 33, Teotihuacan, Mexico. In *Ceramic Ecology Revisited, 1987: The Technology and Socioeconomics of Pottery,* edited by C. C. Kolb, pp. 199-226. BAR International Series 436(i), Oxford.
Sinopoli, C. M.
1991 *Approaches to Archaeological Ceramics.* Plenum Press, New York.
van der Leeuw, S. E.
1976 *Studies in the Technology of Ancient Pottery.* Organization for the Advancement of Pure Research, Amsterdam.
1984 Dust to Dust: A Transformational View of the Ceramic Cycle. In *The Many Dimensions of Pottery: Ceramics in Archaeology and Anthropology,* edited by S. E. van der Leeuw and A. C. Pritchard, pp. 707-773. University of Amsterdam, Amsterdam.

THE FIRES WITHOUT AND THE FIRES WITHIN: EVIDENCE FOR CERAMIC PRODUCTION FACILITIES AT THE LATE CLASSIC SITE OF LA SIERRA, NACO VALLEY, NORTHWESERN HONDURAS, AND IN ITS ENVIRONS

Patricia A. Urban, Naco Valley Archaeological Project, Kenyon College, Anthropology-Sociology Department, Gambier, OH 43022

E. Christian Wells, Oberlin College and Naco Valley Archaeological Project

Marne T. Ausec, Anthropology, University of Massachusetts, Amherst, MA 01003

THE NACO VALLEY AND THE SITE OF LA SIERRA

The Naco Valley lies along the middle Chamelecon River, northwestern Honduras (Fig. 1). The valley encompasses 96 square kilometers of largely flat land between 100 and 200 m above sea level. Archaeological work in the valley began in 1975, under the direction of John Henderson (Cornell; Henderson et al. 1979). Subsequently work was directed by Patricia Urban (1978-79), and Urban and Edward Schortman (1988-present). Survey has recovered more than 300 sites in the valley proper (Fig. 2 shows sites mentioned in the text) and an additional 61 from adjacent areas, with occupation beginning in the Middle Preclassic (800-400 B.C.) and running to the Spanish conquest (A.D. 1531). The best-represented period of occupation is the Late Classic (A.D. 600-950) and Terminal Classic (A.D. 950-1050). Approximately 80% of surface-visible construction in the valley can be dated to this period. During the Late Classic, the valley was united into a single polity the regional capital of which was the site of La Sierra (Figs. 3 and 6). La Sierra contains 485 surface-visible platforms; excavation suggests that as many as one-third that number lie hidden beneath overburden. The site is divided into a site core, a dense North Group, and a less densely settled South Group, with more

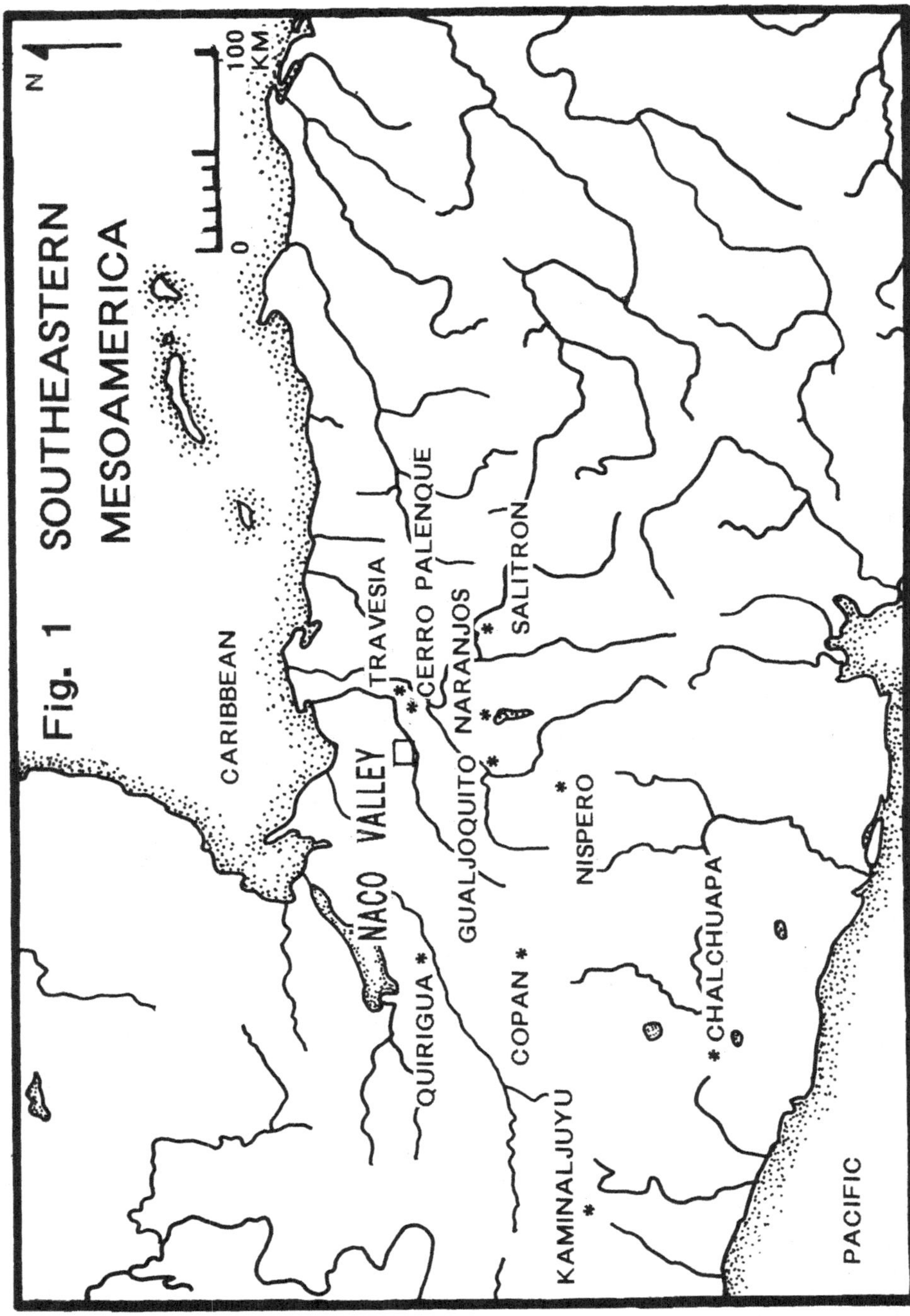

Fig. 1 SOUTHEASTERN MESOAMERICA

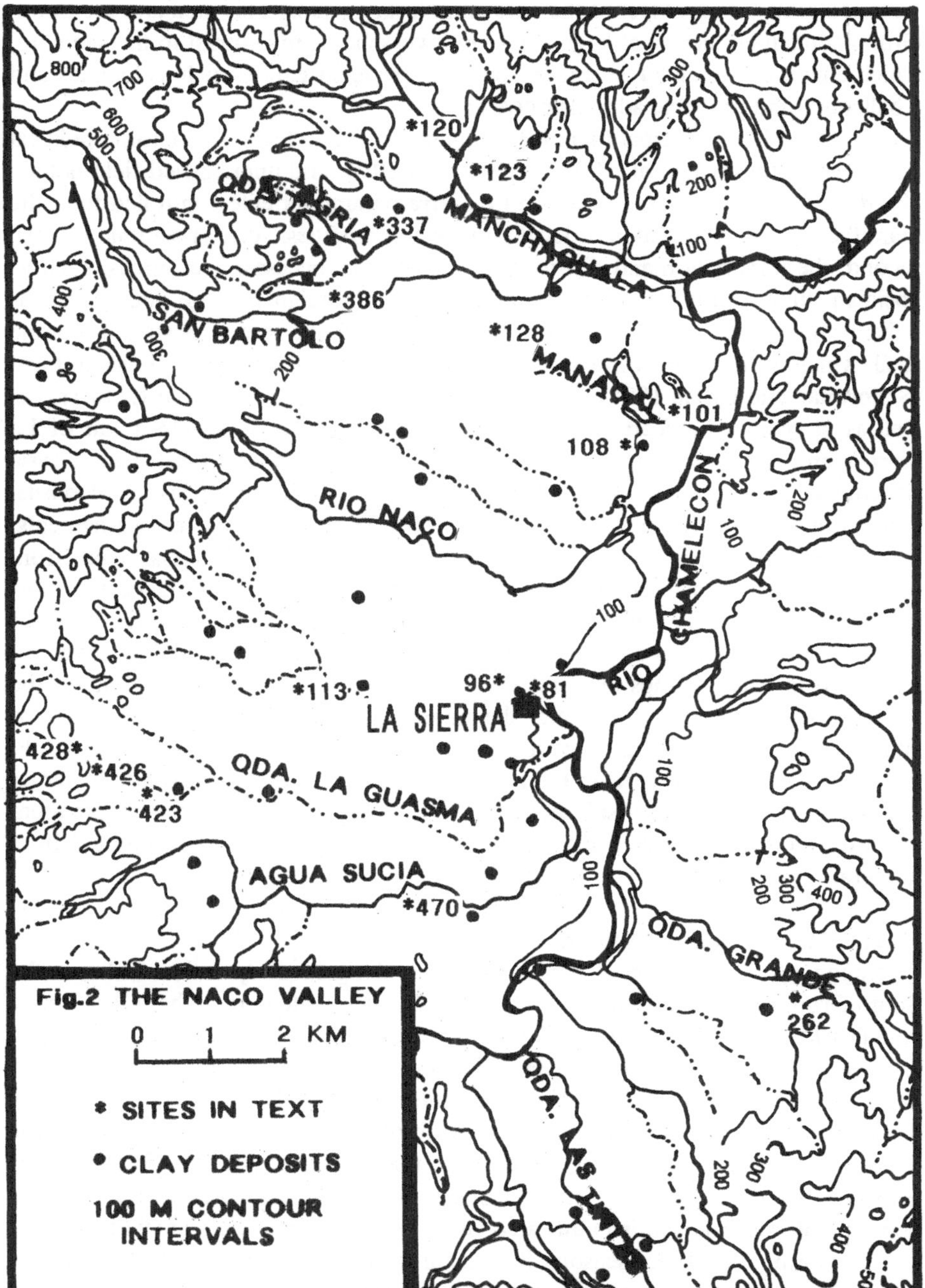

Fig.2 THE NACO VALLEY

scattered groups to the west of the core and North Group. This area and the South Group contain domiciles, storage areas, and workshops, in contrast to the site core, which encompasses elite residences, special purpose structures such as temples and a ballcourt, and probable administrative areas.

One of the principal foci of investigation is the valley's Late Classic political economy (Schortman et al. 1991a, 1991b, and 1994, and unpublished field notes), i.e., determining the degree to which La Sierra paramount elite controlled first, the manufacture and distribution of locally-made commodities, such as ceramic vessels, second, the importation and distribution of non-local necessary resources, such as obsidian, third, the importation and distribution of foreign prestige goods such as polychrome pottery, and fourth, the importation of exotic resources such as marine shell which were then worked and exported from the valley in finished form (Schortman and Urban 1994). To this end, we have excavated within La Sierra, as well as at sites in La Sierra's near periphery (territory lying within 1 km of La Sierra), and sites in the capital's rural hinterland. We have tested 339 Late Classic structures clearing ca. 10700 square meters of Late Classic deposits. Approximately 680 000 pot sherds, roughly 20% of all recovered vessel fragments, have been analyzed, as well as at least 4000 pieces of *bajareque* (clay daub from wattle and daub construction), and over 12000 other ceramic artifacts, the most important of which for this paper are the figurines, ocarinas, and whistles (referred to as FWOs below; more than 4500 studied) and molds for FWOs (over 80 recovered to date), stamps (1), and vessels (4). Project nomenclature follows the operation-suboperation-lot system: the numerical operation designation refers to either a site or a defined area within La Sierra proper and is abbreviated as OP, e.g., OP 43; the following letter refers to a specific excavation area within the larger operation, known as the suboperation (OP 43B). Within the suboperation, excavator-defined provenience units are designated as lots, and represented by a numeral.

CERAMIC WORKSHOPS AT LA SIERRA

Operation 43: The portion of the La Sierra site designated Operation 43 lies in the South Group (Fig. 3), just south of the main La Sierra site core. The OP 43 zone contains 11 surface-visible structures on its west side, a large depression some 25 m E-W by 30 m N-S (at most 2 m deep) in its center, and to the southeast a second depression (ca. 20 m E-W by 30 m N-S, and maximally 1.5 m deep) bordered by disturbed structures. Immediately to the north of OP 43 are OPs 39 and 41; these 2 zones have 3 and 6 structures respectively, that border on the OP 43 depression. Between the depressions no structures are visible on the surface, but there is considerable ceramic debris in this area, disturbed by human and animal activity.

During the 1995 season, 3 structures on the northwest border of the depression were tested, as was one on the southwest (Fig. 4 shows the '95

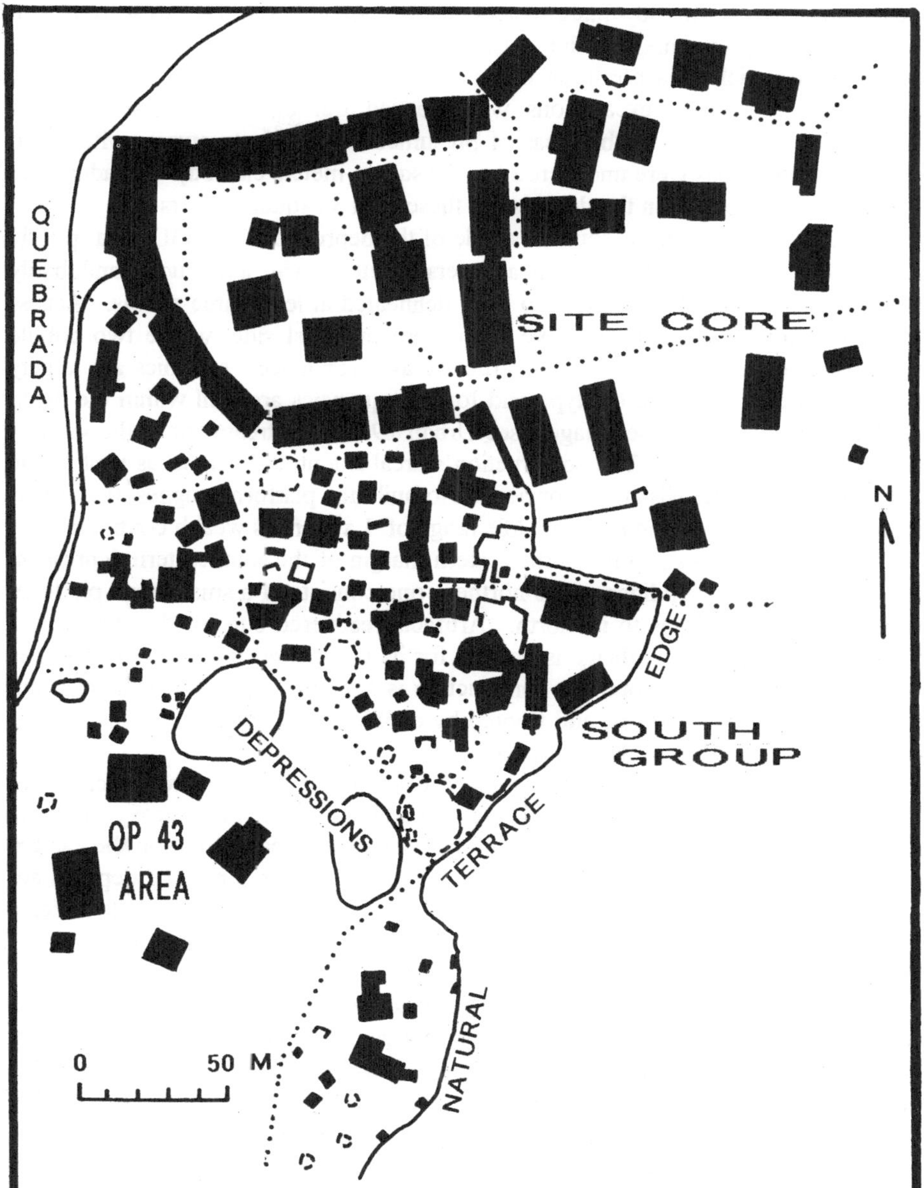

Fig. 3 LA SIERRA SITE CORE AND SOUTH GROUP

excavation plan), with 130 square meters cleared. In addition, test pits were placed across the depression, on the east side of the depression, and in the ceramic-debris-rich area northeast of this borrow pit. In all, 35 1x1m pits were completed. In 1996, excavations were carried out on the east side of the southeastern depression in the areas of presumed disturbed structures, where two distinct constructions were uncovered (ca. 30 square meters cleared). An additional 15 test pits were placed in the depression; these total 33 square meters.

The structure on the southwest side of the depression (OP 43D) is a poorly preserved, two level platform with a superstructure. The ambiguous and badly preserved summit features appear to have delineated at least three rooms. Access to the construction appears to have been from the west side, where two cobble steps were found. We interpret the structure as a residence: ceramics are largely domestic in character, and the type and form frequencies are well within the norm for La Sierra domestic assemblages (see Urban 1993 for a treatment of the valley's type-variety-mode taxa). The construction's areal extent and height, as well as the amount of both materials and labor needed to build the platform, suggest that it was inhabited by people falling in the middle range of La Sierra's wealth continuum.

The two constructions on the eastern margin of the southeastern depression are likewise in poor condition. The northern one consists of a small stone platform with surface-level rooms to its north. Artifacts recovered suggest that this was a workshop area: the assemblage is notable for its worked obsidian, and soft stones used for pigment. The southren construction is a 1.4x1.7 m basin built from cut tuff blocks and river cobbles, and floored by cobbles. The function of this basin is unclear, but it may be for clay levigation.

The southeastern depression tests revealed an ancient borrow pit (ca. 18 m E-W by 20 m N-S, and 1.5 m deep) dug into a clay deposit with a heavy sand and gravel admixture. The eastern pit margin has two ash deposits, thin on the eastern edge of the ancient pit, but up to 1 m thick within the depression. The deposits are not clearly associated with final phase or earlier constructions, but indicate extensive burning in the area. No evidence of kilns, however, appeared in any of these eastern probes.

The tests across the western depression showed that it, too, is an ancient pit, with edges covered by Late Classic midden material. In contrast to the southeastern depression, where a basal clay was encountered, with lenses of a darker clay in the pit fill, three types of clay are found in the western depression: black; brown, but similar in texture and inclusions to the black clay; and a tan, highly micaceous earth. We have not yet completed testing of these clays, but macroscopically, the black and brown appear similar to clays commonly used for ceramic vessels, while the tan micaceous level is similar to materials used, primarily, for figurines, whistles, ocarinas, and stamps.

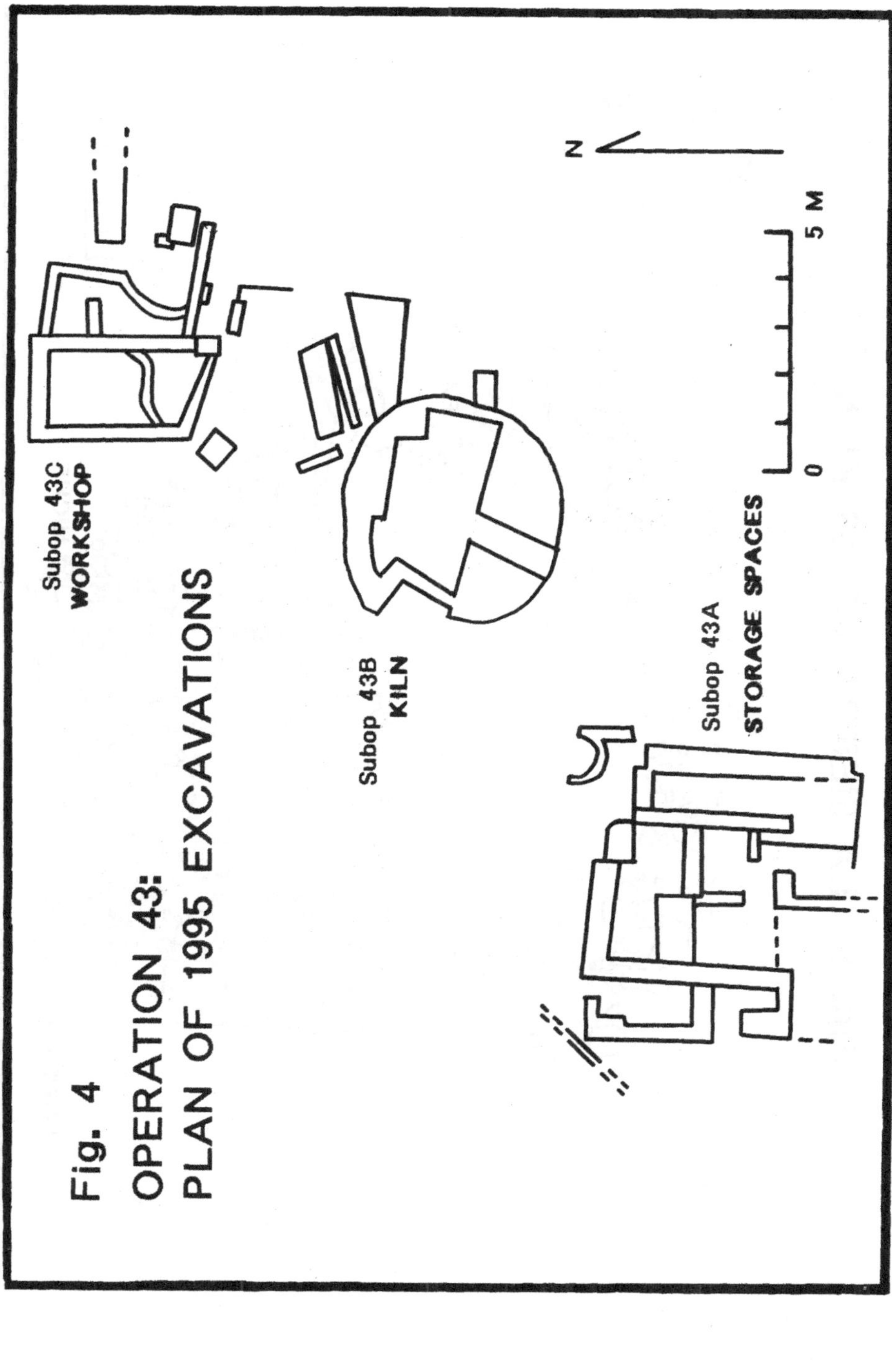

Fig. 4
OPERATION 43:
PLAN OF 1995 EXCAVATIONS

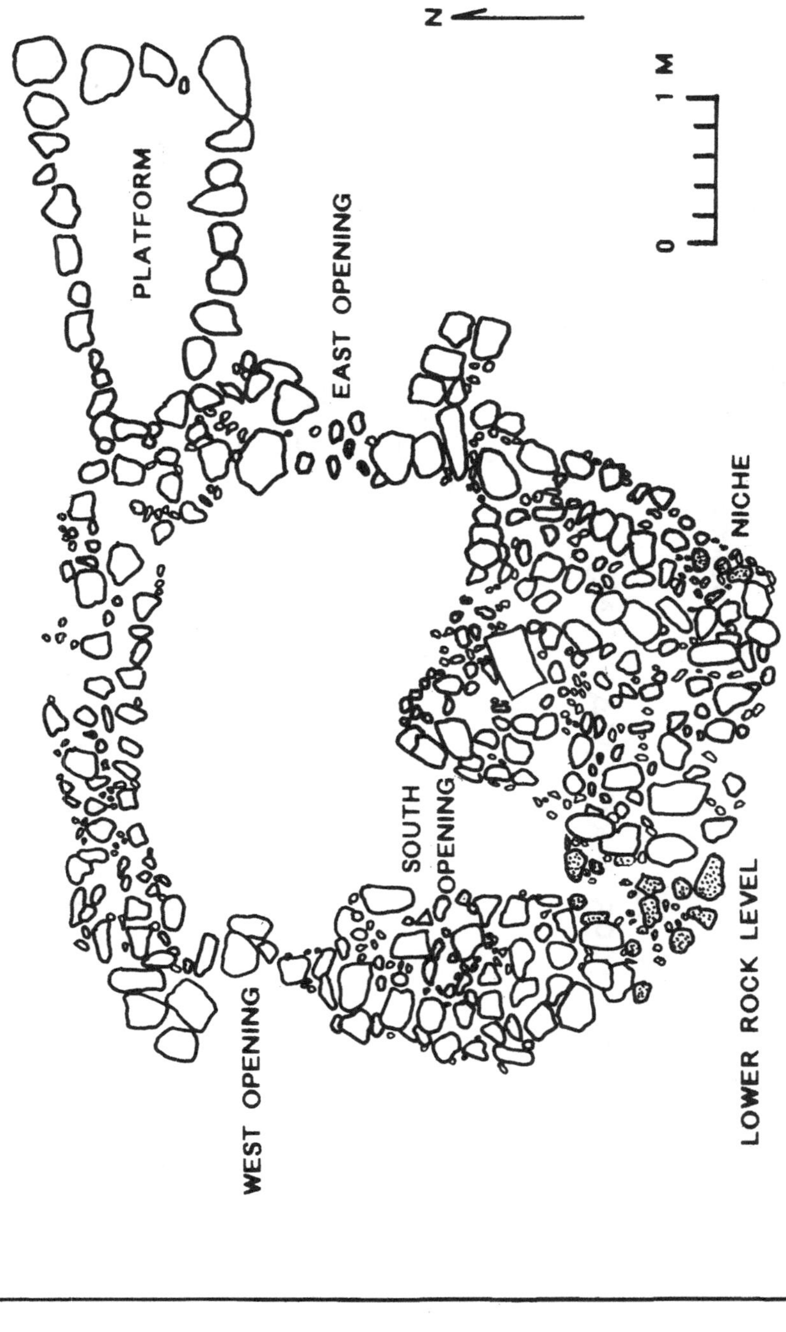

Fig. 5 LA SIERRA OP 43 KILN

Test pits on the ridge between the two depressions revealed an extensive but shallow midden deposit, generally 0.6-1.0 m thick. In contrast, the tests on the northeast margin of the western depression revealed much deeper deposits: due to time constraints, we did not reach the bottom of these cultural strata, but the remains in the central test were undiminished at 2 m below ground surface.

In all test pits, vessel fragments are the most common material. Type frequencies for test pit ceramics lie within the normal Late Classic range: utilitarian types predominate (52.4-69.2% of the assemblage), but for pits within the western depression proper frequencies of red-on-natural taxa and imported Ulua polychromes are about half the La Sierra norm.

The structures to the northwest of the western depression show three different types of constructions. The southernmost structure (OP 43A) is a complex of 7 small rooms oriented N-S. The largest space is 2 m E-W by 3 m N-S; the smallest is a mere 1 x 2 m. In size and construction these features are similar to storage facilities found elsewhere in the Naco Valley, and documented in the Middle Ulua drainage as well (Schortman and Urban 1995), an interpretation supported by the large quantity of ceramic debris found in the spaces. In most similar storage areas, the ceramics are dominated by jars; here, however, bowl forms are more prominent than expected, double their usual frequency, while jars are concomitantly reduced.

The northernmost platform (43C) is less easy to interpret. In its final form, it consisted of at least one retaining wall and a 3.3x2.2 m room oriented to and accessed from the east. Abutting the room are three semicircular features, relatively uniform in size with maximal diameters of 1.4 m, paved with cobbles, and with each open to a different cardinal direction. South of these semicircles, and using one of their walls in its construction, is a 1x3 m room. The open rooms and recovered artifacts suggest both domestic and work activities: jars dominate the ceramics; 15 metates fragments were found; pigment stones and quartzite were found in association with several paired manos and metates; and a figurine mold was encountered which matches with a figurine found outside the construction.

The central of the three northwestern structures to be investigated (OP 43B) proved to be a circular construct 5.2 m in diameter (Fig. 5 is a detailed plan of this construction). The feature's walls vary from 0.23 m thickness on the north, to 2.38 m on the south. This southern section contains a distinct open area or channel, 2.38 m long, 0.78 m across, and 0.24 m deep, which we interpret as a vent. The vent is stone-paved on its southern half, but earthen on the north. The northern half of this circular feature is an open, earth-floored area bounded by narrow stone walls. On the east and west sides there are small gaps in the walls, roughly 1.2 m wide and 0.15 m deep, vents or possible stoking holes. In our tentative reconstruction, the firebox was in the northern half of the feature, with items to be fired placed on the stone-paved southern half. Heat would have traveled up and across the vessels, venting on the south.

On the northeast of the circular feature is a small raised platform defined by parallel cobble wall foundations; it is 2.79 m long E-W by 1.6 m, standing only 0.35 m high. The function of this area is unknown, but we speculate that it may have served as a fuel storage area. The entire kiln-platform complex was built on a pre-existing slope. The small platform seems to precede the circular feature in the construction sequence

The OP 43 zone was the locus for activity throughout La Sierra's history: the analyzed ceramics contain Middle Preclassic and Early Classic materials, in addition to the predominant Late Classic pottery and a small but consistent Early Postclassic component (including Plumbate and the more common local Early Postclassic types). The area, then, continued to be inhabited long after the decline and general abandonment of the La Sierra central group. Further evidence of long-term occupation is the ashy soil stratum which underlies all tested construction. As a rule, it begins 65 cm below ground surface; we never reached its final depth, but it is at least 0.35 m deep in most tests. Conservatively measured, the ash covers 40 m N-S by 25 m E-W, conforming to the ancient slope, but not present in the borrow pit.

The OP 43 area shows us that pottery manufacturing was integrated into residential areas in La Sierra's South Group. Here, with the circular feature, is clear evidence of production. The depression and its clay content also support the actual construction of vessels in this area. If the kiln had not been uncovered, it would be difficult to see ceramic manufacturing in this spot: the assemblage of discarded ceramics does not, overall, have any markedly unusual proportions of either types or forms; the tools located differ little from those in other parts of La Sierra, only showing slightly higher numbers of scrapers and smoothers made from sherds; and there are no clear "wasters" or damaged vessels. The only non-structural line of evidence pointing towards manufacturing is the non-house *bajareque* (daub), which replicates characteristics found at another kiln in the zone designated OP 31.

Operation 31: During the 1990 season, extensive investigations were carried out in the North Group, lying immediately north of the Site Core across a now-dry *quebrada* (stream course; Fig. 6). Due to the dense construction in this cluster, we have divided the area into operation areas, each operation conforming to the surface-apparent coherency of patio groups. The structures on the northern edge of the North Group are included in Operation 31. Here we cleared 500 m2 in a zone bounded on the north and east by a drop to a lower terrace, and on the south and west by modern roads which, at least on the southern edge of the operation area, destroyed several platforms. The zone contains 10 surface-visible structures,

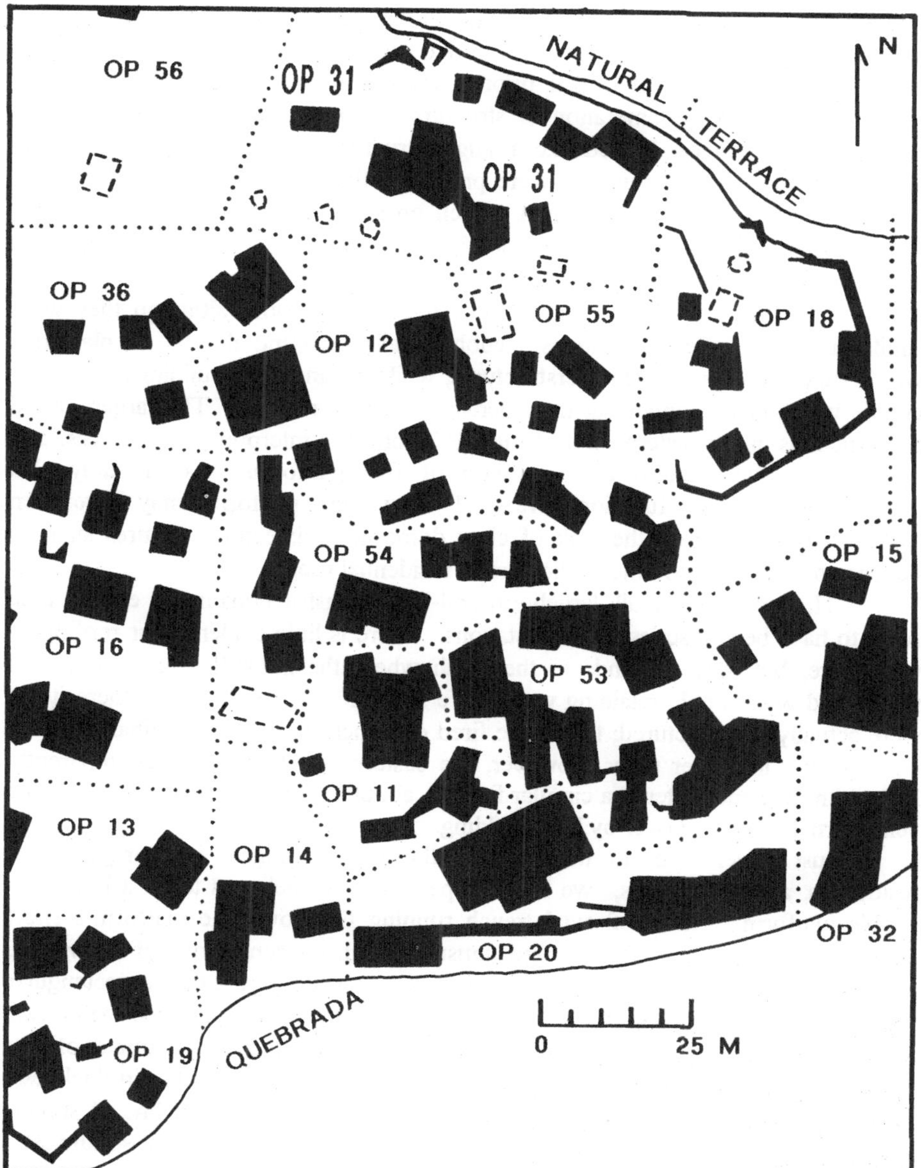

Fig. 6 LA SIERRA NORTH GROUP

and a depression which we initially attributed to modern road building. Excavations showed that the surface view was a simplified one: 5 sub-surface buildings were encountered in an area that looked, prior to excavation, like an open patio area; and to the west of this agglomeration of structures we unearthed 5 cobble platforms which, on the surface, looked like a single structure (see Fig. 7, the excavation plan). In total, 18 constructions were probed in the OP 31 area. Test pits were also dug in and near the depression, which proved to have been dug into dense clay. Construction in the OP 36 area south of the depression conformed to the pit's ancient slope, supporting its precolumbian origins.

All of the excavated buildings differed in final form, although they were similar in construction techniques: cobble walls surround the fill of platforms which supported perishable superstructures, while ground-level buildings have only cobble wall footers supporting their perishable upper reaches. The larger, raised constructions in the eastern part of the OP 31 area we interpret as residences, and one part of the western agglomeration of five platforms was probably also residential in its main function. Most of the western platforms may have been storage areas, based on the lack of clear domestic architectural features such as benches or shelves, and other indicators of residential usage.

The remaining small, largely ground-level, constructions in the eastern area seem to have been a series of work stations. There is little evidence for roofing or walls here, but trash abounds to the north where the naturally steep slope was moderated with cobble retaining walls. We see this zone as the area where items were actually manufactured; they were fired elsewhere in the zone's final phase of occupation. In earlier times, however, the eastern structural complex apparently had a kiln: we encountered a circular feature, about 1 m across and with a possible vent, partially covered over by later building.

Just under 4 m to the west of the total aggregate of storage areas, residences, and workplaces, we had mapped what we believed to be a long, low cobble platform. An initial axial trench running E-W over the platform's short dimension revealed curved walls and unusually high concentrations of *bajareque*, as well as very large quantities of pot sherds. Clearing revealed a rectangular platform 4.4 m long N-S by 6.1 m E-W, constructed and surfaced with river cobbles. It abuts a circular feature to its north which has an overall diameter of 6 m (Figs. 8 and 9). The walls of the circular feature vary from 1.5-2.0 m in thickness, enclosing a space 2.8 m across. The inner circle is cut by a single line of stones, preserved one cobble high and one cobble wide, running northeast-southwest . Both the north and south sides of the circular feature contain formalized, rectangular channels, the northern one lined with cut blocks, the southern with cobbles. There is also an indentation on the southwest side of the feature. The indentation may be a stoking hole for a fire box located in the southern half of the circle. The southern vent would then have been part of the fire box. The northern

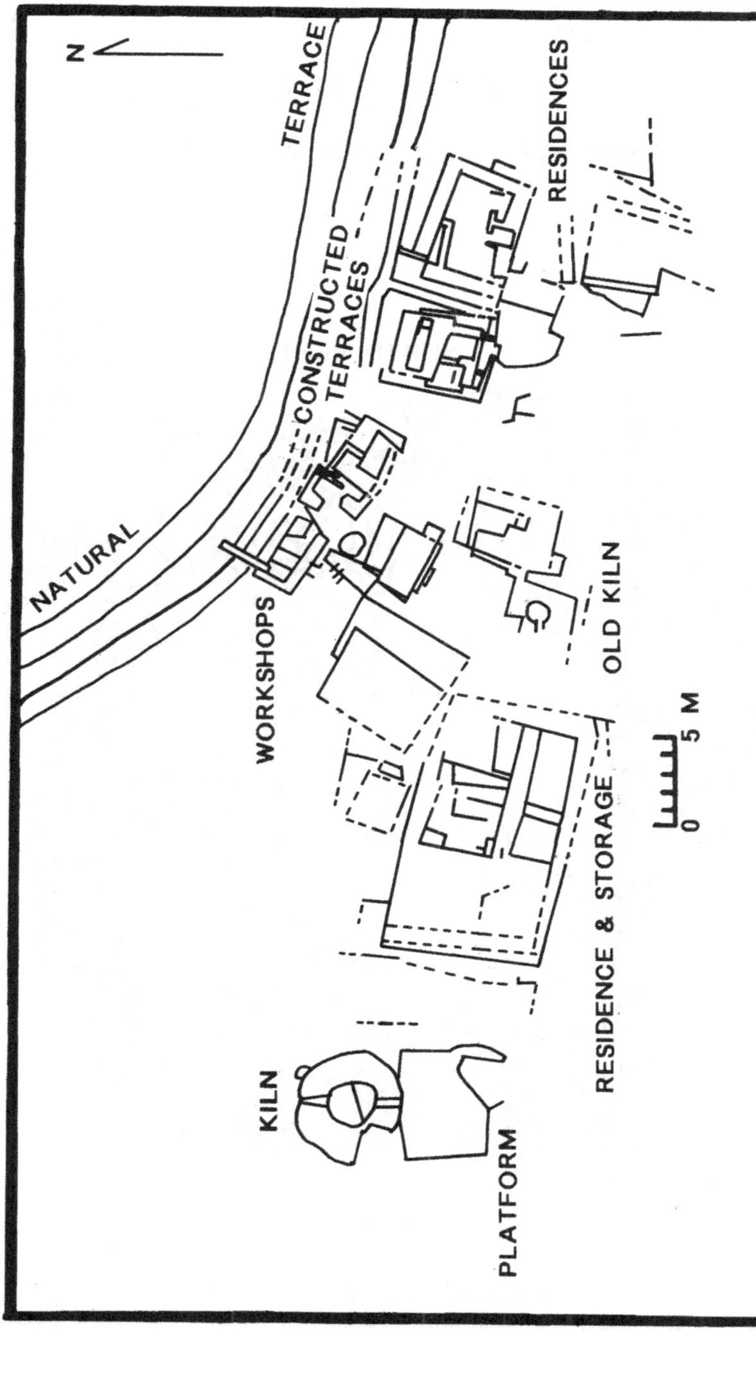

Fig. 7
OPERATION 31 EXCAVATION PLAN

Fig. 8 LA SIERRA OP 31 KILN

WITHOUT SOUTHERN ABUTTED PLATFORM

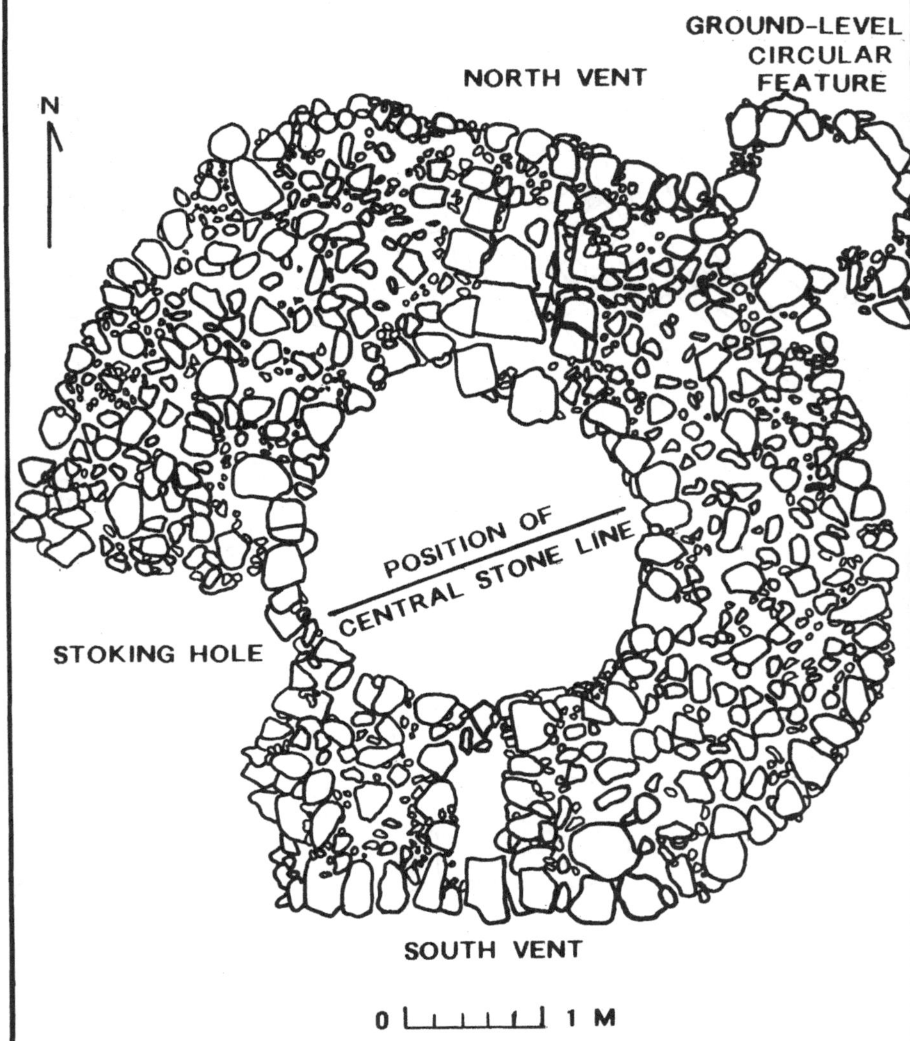

half of the circle may be the actual area for vessel placement, with the vent functioning to provide fresh air or to exhaust unwanted heat or smoke. In all, we see this construct as a circular, two-chamber, downdraft kiln, with a beehive-shaped upper section.

Evidence for the upper part of the kiln comes from the extensive *bajareque* deposits found around and in the circular area. This *bajareque* differs substantially from that used for houses: house daub tends to be dense, with sand tempering and little organic material; the support posts and cross-bars of the wattle framework are straight, are always composed of whole, unsplit branches and saplings, and are selected for uniformity of size; the framework itself is rectilinear, with supports and cross pieces forming right angles; and while some house *bajareque* is hard, apparently accidentally fired, the majority is still soft, easily eroded, and found in small pieces, rarely over 5-7 cm across. In contrast, OP 31 kiln *bajareque* is highly porous, apparently tempered by large quantities of grass and lacking sandy inclusions. The pieces tend to be hard, and some fragments were heated so often and/or to such high temperatures that they show signs of vitrification and melting. The pieces themselves tend to be curved, with exterior surfaces rougher and lumpier than house exteriors. The marks left by the post and stick armature are also curved; in some instances it is possible to see converging stick patterns and lashed stick and post crossings. The framework never joins at right angles, making instead pointed junctures, as would be expected from a beehive-shaped superstructure. Many of the sticks are whole, but there are also numerous split saplings or branches; the incidence of splitting approaches 75%. Splitting no doubt made the wood more pliable for constructing the arches necessary to support the *bajareque*.

Soils in the area are distinctly ashy, with a 0.4 m deep layer of ash underneath the kiln and associated platform, and extending out around it for as far as we cleared (1-2.5 m). Analysis of the ash indicates burning of hardwoods (Schortman et al. 1991). The area also produced vast quantities of ceramics, largely of Late Classic date; we have analyzed over 40 000 fragments from the OP 31 zone, 80% of these from the kiln area, for a total sample of ca. 25%. These items did not show proportions of types or forms that differed in any way from those in presumed non-ceramic-producing zones, nor was there a higher incidence of bubbled or cracked items than we find elsewhere. (During the 1991 and 1992 seasons, all 185 583 analyzed sherds were classified within their type-variety taxa according to variables relating to firing such as bubbling, hairline surface cracks [crazing], fire-clouding and so forth; no deformed items were found. The sample includes excavation loci outside of and within ceramic production zones, from La Sierra, the near and far peripheries.) While the bulk of the ceramics found around the kiln were Late Classic, there was an admixture of later materials from the Early Postclassic. The northern vent contained large sherds that typologically were Early

Postclassic, and this date was supported by a lone C14 assay associated with the final use. These data suggest that the kiln operated until approximately A.D. 1200.

The platform abutting the kiln may have served for wood storage, drying of finished ceramics, or as a staging area for other parts of the production process. A curved area in the southeast corner suggests either an earlier kiln in this spot, or the beginnings of a second kiln that was never finished.

Other Indication of Ceramic Production and/or Firing at La Sierra: While the Op 31 and Op 43 zones show the clearest evidence for formal firing facilities, other parts of La Sierra also have aspects that point to controlled use of fire. Some of these may be domestic ovens, while others may have been used for manufacturing purposes. First, the summit of a raised platform just south of the OP 31 workshop zone, in the OP 36 area (Fig. 6), was reused as a firing/baking facility. Within a rectangular room a round feature was constructed which supported a possible beehive superstructure. The earthen flooring and *bajareque* within the circular feature and on its wall stones had been altered by heat: their color and texture had changed in places, and there was evidence of melting and vitrification in the *bajareque*. Moreover, the daub shared many of the characteristics of the OP 31 materials: porosity, a dearth of aplastic inclusions, and a non-house pattern of stick impressions. Second, just southeast of this reused platform is a rectangular stone basin (northern edge of the OP 12 area), comparable in size to the OP 43 basin, but better made an inset into a cobble pavement.

Third, a structure in the far west of the OP 31 zone also revealed a room remodeled and used as a firing facility. Again, there were signs of thermal alteration, and the *bajareque*, while not unequivocally like that from the larger kiln, shares more characteristics with kiln materials than it does with house remains; this places it in our third category, oven/possible kiln. Fourth, in the southeastern part of the North Group (OP 32) a platform summit had partially melted, glassy rock fragments and *bajareque*, but no other clear evidence of a kiln or oven. Fifth, in the southwestern part of the North Group (OP 19 zone) is a potential oven base--a raised, rectangular feature, covering just over 1 m^2, with unusual quantities of oven-like, if not purely kiln-type, bajareque over and around it. Very heavy sherd deposits were uncovered in the immediate vicinity of the square feature. Sixth, in the west-central part of the North Group (OP 16 area) are three oval, ground-level features, delimited by cobbles. Burned earth and charcoal were found inside. These have no clear kiln or even oven *bajareque*, and we interpret these more as possible cooking pits than manufacturing features. Seventh, and finally, kiln-like daub came from excavations in OP 46 (not included on either map; west of the Site Core, southwest of the North Group), although no clear foundations for a kiln were revealed.

FIRING AND WORKSHOP ZONES OUTSIDE OF LA SIERRA

Investigations outside of La Sierra have revealed situations analogous to all La Sierra firing areas, except the very large formal kiln. At Site 426, in the southwest corner of the valley (Fig. 2), the final phase of a raised platform included some sort of firing facility. The platform supported two summit rooms, the eastern one 1.2x1.5 m, the western 1.75 m on a side. In the western room were burnt earth and many sherds, some of which were thermally altered, along with ash and charcoal flecks. *Bajareque* inside and around the western room has many of the formal characteristics of the OP 31 bajareque. The limited size of the area suggests small-scale firing of vessels, or firing of small items such as figurines; molds for FWOS were found at the site.

In our 1996 work at Site 128, in the northwest portion of the valley, two circular features have been recovered, both parts of final phase, Terminal Classic, construction. This site also has a large depression, which we once believed to have been from modern tile-making (the remains of which are still visible to the north of the site), but which after testing looks like an ancient clay borrow pit such as those found at La Sierra. Soils on the southern pit edge contain some ash and charcoal flecks, and large quantities of sherds. Near the circular features are other deposits of extremely ashy soil. The site also has produced FWO molds, although these come primarily from fill.

Two valley loci present evidence for the production of ceramic items, but not necessarily their firing. At Site 337, in the northwest portion of the valley close by the base of the foothills, a large quantity of molds was found: 29 molds are for the figurative portions of figurines, ocarinas, and whistles, while 3 fragments are for molding small bowls. Neither of the two investigated structures (of three surface-visible platforms) showed any traces of firing facilities, nor did the non-structural areas. In all 120 m^2 were cleared at Site 337, an area which encompasses the two platforms and inter-platform space.

In the south-central portion of the valley, again near the base of the foothills, some evidence for manufacture of small ceramics was found, but again little sign of firing. At Site 470, 2 FWO molds were located as well as sherds with lumps of fine clay adhering to their exteriors. Some of the lumps were merely dried, but a few were fired hard; in several instances, the fired clay covered ancient sherd breaks. These sherds may have been used to seal a small firing facility: excavations revealed a rectangular feature with burned earth in and around it.

Finally, evidence for high temperature and/or long-term heating was found at two valley locations: Sites 423, in the southwest sector; and 262, which lies east of the Chamelecon River. In both sites small amounts of melted and partially vitrified clay, probably *bajareque*, were found, but without other suggestions of firing areas. At the time of writing, artifact analysis for the 1995-96 excavation was not complete; further work may turn up more molds and/or kiln- or oven-type daub.

SEEING CERAMIC WORKSHOPS IN THE NACO VALLEY ARCHAEOLOGICAL RECORD

In the Naco Valley ceramic workshops and kilns are not detectable from ground survey, nor are they accessible through surface collections. Sites that are undisturbed rarely have more than 2 or 3 dozen sherds over their entire surface areas. Looted or plowed sites produce more, but the finds generally come from fill, not final-phase use, and are therefore not useful for assessing the latest occupation. Surface collections never have *bajareque* larger than 3-4 cm in diameter, and even that is uncommon. We have also learned that surface materials are rarely a good guide to sub-surface remains: surface ceramics tend to be body sherds from undecorated utilitarian wares of uncertain date; rare, distinctive material has tended to be from fill, skewing both temporal and functional analyses. Moreover, we have not detected any marked differences in the ceramic assemblages from <u>excavated</u> workshop and non-workshop areas, nor can we identify any complex of tools definitely associated with ceramic working, other than molds. How, then, can we assess the presence or absence of ceramic working at any given valley locus?

The best, and often only, line of evidence is unambiguous kiln construction revealed by excavation. To us, an unambiguous kiln displays the following features: First, it is generally, though not always round--rooms reused as firing facilities retain their square or rectangular proportions. Second, the *bajareque* will be lighter in weight, more porous, and usually thicker than house daub; it will also lack the aplastics common to house materials. Any wattle-marks remaining will differ from house materials in form, placement, size, and shape. Third, alteration of *bajareque*, stones, and/or sherds by heat should be detectable. This alteration generally includes hard-firing, even melting and partial vitrification, of the daub, discoloration, cracking, and occasional melting and vitrification of the rock, and color and texture changes in sherds used as kiln furniture or kiln sealing material. From our refiring studies, we are now able to identify the color changes local types go through as they are heated and reheated, and this has been our clearest key to thermal alteration of ceramic materials. Fourth, large quantities of sherds are generally found around kilns--although this is relative, since the Naco Valley's ancient inhabitants seem to have used vast quantities of ceramics during the Classic period. Fifth, there usually are deposits of at the least ashy, charcoal-flecked soil, if not ash outright.

On the other hand, we do not expect to find the following characteristics. First, although there will be large quantities of sherds, their vessel forms and type frequencies will not depart from the norm for non-workshop areas. Sherds will also not show any preponderance of misfired items: in all parts of the valley we find fire-clouded, bubbled, and crazed items. This could signify almost universal manufacturing of vessels, but in the absence of supporting evidence for this interpretation, we believe that minor surface flaws were not a detriment to a vessel's use, with flawed vessels universally distributed. As an aside, we are no longer seeing spalling as the necessary by-product of bad firing; instead, it may indicate something about how the finished vessel was used (see Skibo 1992 for a stimulating discussion of use-wear on modern vessels). The non-pottery artifacts will not be particularly indicative of ceramic working. Formal scrapers (made by shaping broken pieces of pottery, generally into disks) are found in a wide variety of contexts; their use seems not to be limited to ceramic production zones. Informal scrapers and smoothers, the products of casual use of sherds, may be slightly more common in ceramic workshop areas than other places, but the verdict is still out on this; furthermore, they are difficult to impossible to detect if the lots are not well-preserved. Molds are often encountered in unambiguous firing areas, but we also find molds in places where there are no signs of kilns or ovens. This suggests that molds merely indicate the formation of vessels and FWOs which in some instances were taken to other locales for final finishing and/or firing. We should also note that molds seem to be widely dispersed in Terminal Classic contexts, possibly implying temporal changes in the valley-wide manufacturing system which we cannot yet fully adumbrate. Ceramic producing areas tend to have notable quantities of soft, colored stones, probably used as pigments or simply as "chalk" for sketching designs, but these stones are found in non-workshop areas, too.

Clay sources have not been clear indicators of ceramic manufacturing area. To date, 56 surficial clay deposits have been located in the valley (Connell 1995; Schortman et al. 1991; see Fig. 2). Modern potters indicate that all of these are at least satisfactory for vessels and/or FWO production. In only two places identified as clay sources--La Sierra itself and Site 128--have clear firing facilities been found, while at a third, Site 108, Connell sees vessel production as fairly certain. Furthermore, access to clay may have been a less significant variable than access to firewood. A final point is that in the Naco Valley, social and political factors conditioned workshop locations as well, further confounding attempts to predict where ceramic production might have taken place.

In ambiguous situations, that is, those in which kilns are not readily apparent (such as the three instances of room re-use), we take a cautious approach. If the *bajareque* is definitely of the kiln type, we are inclined to postulate some sort of heat-producing activity. These activities may include actual ceramic firing, but such features may also simply be ovens for cooking. There is an intermediate form of daub--moderately porous, generally with some thermal alteration, but with the wattle framework less clearly making a beehive--that we interpret as oven daub.

When we encounter this sort of *bajareque* without any other evidence for firing, we do not regard the area as a workshop, but instead see it as a potential cooking spot.

In sum, work to date has not produced evidence of ceramic production recoverable from survey alone. Even after excavation, interpretations can be just that--interpretations. Multiple lines of evidence, such as the presence of borrow pits and kiln style daub, large numbers of scrapers and pigment stones, and thermally altered pot sherds, rocks, and or daub, lead to more secure designations of workshops than any one or two categories of evidence alone can.

CONCLUSIONS

In the Late Classic Naco Valley, the best evidence for the complete process of ceramic manufacture comes from the regional capital, La Sierra. Here we have actual kiln facilities, clay deposits, ash layers, and tools. At Site 128 we also have probable firing facilities, clay deposits, and ash layers, while the excavator of Site 108 interprets it as having ceramic production. At Sites 470, 426, and 337, we have evidence for, at the least, forming of ceramic items; in the case of 426, firing is likely. We are currently trying to assess more exactly the temporal aspects of ceramic manufacturing. The large kiln at La Sierra seems to have been maintained and used into the Postclassic, while the Op 43 kiln may have fallen into disuse before that era. It also seems to us that ceramic production was widely dispersed in the valley during the Terminal Classic. Whether this represents a failure of control on the part of elites at the regional capital, resistance to control of production on the part of rural populations, or some other set of factors, we do not know at this time

ACKNOWLEDGMENTS

The work discussed here was carried out by teams consisting of student and staff supervisors and local excavators. We wish to thank in particular Laura Aldrete, Christopher Attarian, Ellen Bell, Johanna Bell, Susan Buchmueller, Benjamin Carter, Lisa Collins, Samuel Connell, John Dietler, Jennifer Ehret, Nicholas Gevock, Hope Henderson, Andrew Kindon, Michael Kneppler, Krista Miller, Beverly Shade, Lavinia True, and Steven Yates for their able supervision during the 1990, 1992, 1995, and 1996 seasons. Sylvia Smith Duggan (1988), Marne Ausec (1990, 1995-6) and Ellen Bell (1992) have been Laboratory Directors. Their principal assistants have been Urban and Edward Schortman. Studies of *bajareque* are conducted by Ausec and Urban, who are also responsible for refiring experiments. Ausec and E. Brown carried out studies of modern ovens; the kiln/oven reconstruction program in 1996 is supervised by Ben Carter.

Research in the Naco Valley has been funded by Fulbright Fellowships held by Urban ('88) and Schortman ('95), as well as by grants from the following agencies: NSF Research Experiences for Undergraduates (1988, 1990, 1992, 1995-96); NSF research (BNS 8919272, BNS 9022247, BNS 9121386,

SBR-9407751); National Geographic Society (4208-89); the National Endowment for the Humanities (RO-21897-89), the Wenner-Gren Foundation (1988, 1992) and a grant from the now unfortunately discontinued Margaret Cullinan-Wray fund (1992), administered by the American Anthropological Association. The Instituto Hondureno de Antropologia e Historia (IHAH) has extended us every courtesy throughout our years in Honduras, and we are particularly thankful for the assistance of Lic. Juan Alberto Duron who is in charge of the northern region, in which administrative division the valley falls.

This paper is dedicated to the people of the Naco Valley, ancient and modern, without whom there would be no understanding of the past, or past to understand, and in particular to Jorge (Zorro) Bueso. We are solely responsible for any errors or omissions herein, with principal culpability on the shoulders of the senior author.

REFERENCES

Connell, Samuel

1995 *Clay sources and Ceramic Production in the Naco Valley, Northwest Honduras.* MA thesis, Department of Anthropology, UCLA.

Nelson, Ben, ed.

1985 *Decoding Prehistoric Ceramics.* Southern Illinois University Press, Carbondale.

Rice, Prudence

1987 *Pottery Analysis.* University of Chicago Press, Chicago.

Schortman, Edward, and Patricia Urban

1991a *Sociopolitical Hierarchy and Craft Production: The Economic Bases of Elite Power in a Southeast Mesoamerican Polity.* Unpublished ms., Kenyon College and IHAH, Tegucigalpa, Honduras.

1991b *Sociopolitical Hierarchy and Craft Production: The Economic Bases of Elite Power in a Southeast Mesoamerican Polity, Part II.* Unpublished ms., Kenyon College and IHAH, Tegucigalpa, Honduras.

1994 *Sociopolitical Hierarchy and Craft Production: The Economic Bases of Elite Power in a Southeast Mesoamerican Polity, Part III--The 1992 Season of the Naco Valley Archaeological Project.* Unpublished ms., Kenyon College and IHAH, Tegucigalpa, Honduras.

1995 Late Classic Society in the Rio Ulua Drainage, Honduras. *Journal of Field Archaeology* 22(4):439-457.

Skibo, James
1992 *Pottery Function: A Use-Alteration Perspective.* Plenum Press, New York.

Urban, Patricia
1986a *Systems of Settlement in the Pre-Columbian Naco Valley, Northwestern Honduras.* University Microfilms, Ann Arbor.
1986b Pre-Columbian Settlement in the Naco Valley, Northwestern Honduras. In *The Southeast Maya Periphery.* P. Urban and E. Schortman, eds. pp 275-295. University of Texas Press, Austin.
1993 Naco Valley. In *Pottery of Prehistoric Honduras.* J. Henderson and M. Beaudry-Corbett, eds. pp 30-64. UCLA Institute of Archaeology, Los Angeles.

Urban, Patricia, and Edward Schortman
1995 *Preliminary Report on the 1995 Field Season of the Naco Valley Archaeological Project, January through May, 1995.* Unpublished ms., Kenyon College and IHAH, Tegucigalpa, Honduras,

Urban, Patricia, Edward Schortman, Sylvia Smith, Julia Miller, L. Theodore Neff, Lavinia True, Marne Ausec, Sonia Kane, Heather Mahan, Matthew Dall, Christopher Eaton, David Schafer, Lisa Collins, Patricia Reed, Patricia Whooley, Nancy Condon, and Richard Goebels
1988 *Socio-political Developments in Northwestern Honduras: The 1988 Naco Valley Archaeological Project.* Unpublished ms., Kenyon College and IHAH, Tegucigalpa, Honduras.

Wells, Christian
1995 *A Preliminary Assessment of the Organization of Ceramic Production at La Sierra, Cortes, Honduras: Evidence from the South Group.* Honors Thesis, Anthropology, Oberlin College.

BRICK AND TILE MAKING IN SPANISH CALIFORNIA WITH RELATED OLD AND NEW WORLD EXAMPLES

Julia G. Costello, Foothill Resources Ltd., PO Box 288, Mokelumne Hill, CA 95245

ABSTRACT

The first kiln in Spanish California to make floor and roof tiles for adobe buildings was constructed at Mission San Antonio by 1781. Archaeological excavation of this facility was conducted in 1976-77. Interpretation of the kiln structure and artifacts were enhanced by studying both Roman precursors of the brick kiln and modern brickmaking processes in Tejar, Guatemala. Similarities between the surviving Spanish Colonial technology and Tejar and that of Mission San Antonio were particularly strong. The lower levels of technological proficiency exhibited at the California kiln site are attributed to poorer clays and to the short (50-year) life span of the colony. The tenuous adoption and mastery of tile making by the native peoples appears to reflect their merely superficial acceptance of Hispanic culture in general.

INTRODUCTION

Among the architectural achievements of Roman civilization was the mass production of fired, construction tiles: flat bricks for paving, walls, and pillars, and curved varieties for roofing. This Mediterranean tile industry—along with an abundance of other European cultural elements—was carried to the New World by the Spanish and adopted by the local populations. The first brick-and-tile kiln to be built in Alta (Upper) California was at Mission San Antonio de Pádua in 1781.

The kiln site represents one of the few mission activities that is well preserved in the archaeological record. Durable and ubiquitous, the distinctive red *tejas* (roof tiles) and *ladrillos* (floor and construction tiles) are characteristic of all mission sites in Alta California. Indeed, the manufacturing of adobes, bricks, and tiles ranked just below the basic industries of agriculture, stock raising, and weaving crafts in the nature of its pervasiveness in mission life and the number of persons employed in its operations (Tays 1941:426). In spite of this, contemporary historic documents are nearly void of technological details on brickmaking activities. Interpretation of the excavated kiln site was therefore assisted through comparison with examples from Roman antecedents and from an operating brickmaking complex in Tejar, Guatemala. At this latter site, resulting brick and tiles appear little changed from those found at Roman and Spanish California sites.

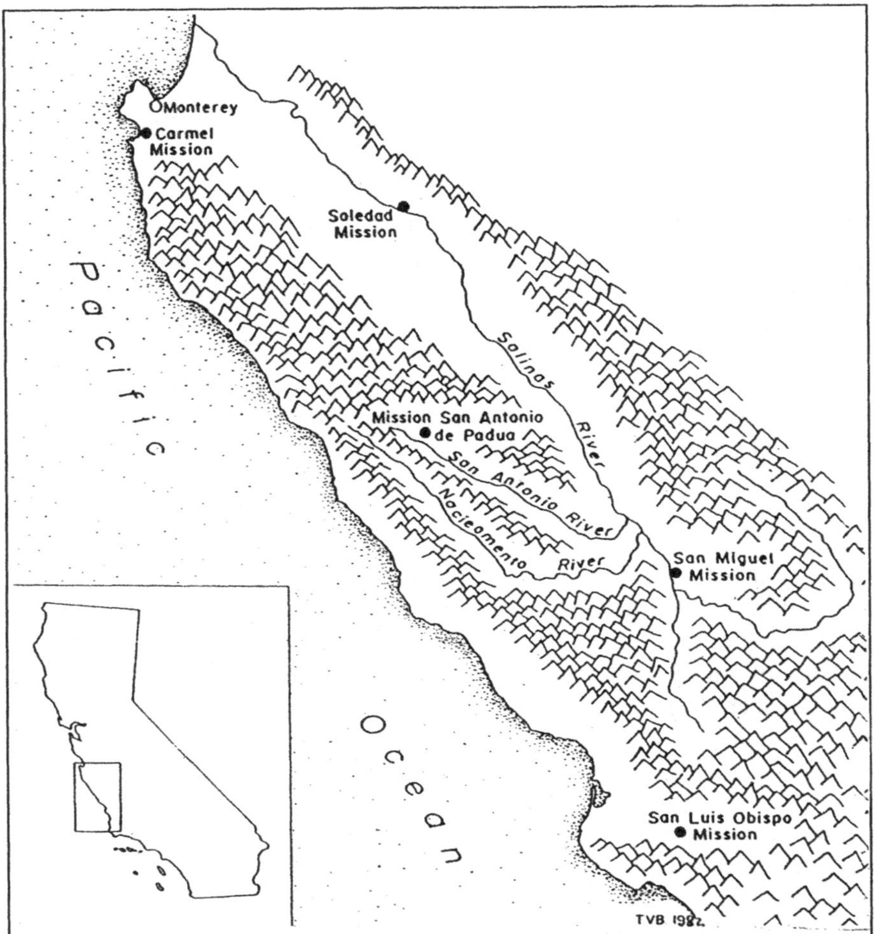

Figure 1. Location of Mission San Antonio in California.

BRICK AND TILE MAKING: FROM THE OLD WORLD TO THE NEW

Fired tiles were being manufactured by Babylonians in the 4th millennium B.C., brick kilns found in Persia are dated to the 1st millennium B.C., and by the 7th century B.C. fired roof tiles were in common use in Ancient Greece. In the earliest years, the most extensive use of tiles was as roofing material and for water channels (Webster 1979:285). As floor coverings, rectangular or square bricks were commonly laid flat or set on edge to form a herringbone pattern. By the third century B.C., tiles were being employed in a wide range of functions requiring several shapes and sizes. Circular bricks with a central hole were used in columns, wedge-shaped bricks formed arch *voussoirs* ("key" tiles holding the freestanding span), square bricks were scored to break into triangles, and layers of bricks were used as bonding courses in masonry work (Wilson 1979:11). Decorative tiles are found in 6th-century B.C. Sicily and by the 4th century B.C. Greeks were experimenting with tiles either pegged or mortared onto walls, tiles shaped to produce columns, capitals, cornices, molded plaques, freestanding ornament and statuary attached to buildings (Blagg 1979:267).

Roman inheritors of this brickmaking tradition incorporated it into their unparalleled development of building construction techniques. Fired tiles were widely used in the construction of both public and private buildings, transforming the face of major cities and identifying Roman constructions in the far reaches of the Empire. Tiles were found in columns, arches, floors, roofs, chimneys, and water systems of the cities and, through hypocaust construction, heated the floors, walls, and ceilings of Roman bathhouses. Fired tiles also provided aesthetic and informative decorations on public, private, and religious structures. Like most of Roman industry, brickmaking had a semi-specialized, or craft, character to its operations (Cram and Fulford 1979:208).

Hispania, greatly influenced by Roman culture and architecture, was said to have become "the most Roman of Roman colonies" (Newcomb 1925:89-90). This influence cannot be overstated, for Roman culture served as the basis of subsequent Spanish civilization. Significant influence, however, also resulted from the Moorish occupation of Spain beginning in the 8th century A.D. and enduring for almost 800 years. Educated and talented craftsmen, the Moors were not constructors but decorators. They adopted Roman engineering and building principles, and added lavish interior decorations, plain exterior wall surfaces, and the Mediterranean courtyard (Newcomb 1925:90-92).

All the technological information and architectural heritage of the Old World was brought to the New by the Spanish. As tile was integral to construction, brickmakers were present on the second voyage of Columbus and thousands of *ladrillos* and *tejas* were recorded as being sent from Seville to the Indies during the first 30 years of occupation (Lister and Lister 1977:55-84). Local tile works and kilns were quickly established in the New World colonies.

Sites of Roman Brick and Tile Kilns

The absence of a prior Hispanic brickmaking technology, and the dramatic and wholesale adoption of Roman culture and industries by the Iberians, argues for a straightforward transplanting of Roman tile making processes. Roman kiln excavations, therefore, can provide information on construction and processing techniques accessible to and adopted by Roman Spain. The most comprehensive discussion of Roman tile making is probably McWhirr's (ed.) *Roman Brick and Tile: Studies in Manufacture, Distribution and Use in the Western Empire* (1979a). Included in this volume are descriptions of Roman period brick-and-tile kilns from Italy, Sicily, and Great Britain as well as reconstructions of kiln operations. Most of the reported tile kilns are square or rectangular, with firing chambers ranging from about 5 m^2 to about 23 m^2. Kiln "plants," with rows of 7 to 10 kilns constructed in a line, have been identified in England and Germany where they are associated with garrisoned Roman legions (McWhirr 1979b:187-198; Sölter 1970). Although the exception, oval tile kilns have been identified in Roman Wales (McWhirr 1979b:186) and modern Sicily (di Caprio 1979:91).

Kiln construction materials include flat tiles, green (unfired) tiles, waster material, adobe blocks, and local stone. Clay was commonly used as mortar and to plaster the walls and floors. Many of the tile kilns were built into hillsides or included a semi-subterranean combustion chamber. This was a simpler structural design than an entirely freestanding kiln and had the added advantages of insulating the lower kiln chamber and of allowing the firing chamber to be loaded and unloaded at ground level. Where the upper chambers of kilns have been preserved, there are no signs of entryways to the firing chamber or of any kind of roofing material. It is presumed that these ancient Roman kilns, like many modern ones, were covered at each firing with temporary material such as kiln wasters or earth (di Caprio 1979:91; McWhirr 1979b:99).

The vast majority of Roman brick and tile kilns identified by McWhirr (1979a) were double chambered. Cross walls in the lower combustion chamber supported the floor of the superimposed firing chamber. Flue holes through this floor promoted circulation of hot gases. Combustion chambers typically had a recessed trough for burning fuel which ran the length of the kiln from the firemouth to the back.

Enough examples also exist to identify single-chambered kilns as among the kiln types available to the Roman tile makers. These have at least partially permanent walls and a recessed central flue, but lack a separation between the combustion and firing chambers; they are referred to as modified "clamp-type" kilns. ("Clamp" kilns have no permanent structures, and the term is typically used to describe a pit or bonfire firing; a clamp-type kiln has an aboveground structure that is built anew at each firing from the green bricks that constitute a load.) Nine of the 51 Roman kilns identified in Britain were modified clamp-type kilns. They were mid-sized, with firing chambers of about 11 m^2.

It is these single-chambered, modified clamp-type kilns—with only the lower walls permanent—that seem to have been most prevalent in the New World. As we shall see in the following sections, both the Spanish kilns at Tejar and the kiln at Mission San Antonio were of this single-chambered variety.

Traditional Tile Making in the Modern World

In ethnoarchaeology, observation of actual cultural situations are used to help interpret the remains of similar activities in the past. In these studies, modern descendants of past groups of people are the most likely to exhibit continuities of techniques and customs (Schiffer 1978:231-232). Although study of the traditional tile making process in modern Spain would provide excellent comparisons with examples from historic New Spain, there is virtually no information available on this technology from the mother country. Modern Italian kiln sites, direct descendants of Roman traditions, are mentioned sporadically throughout the McWhirr volume (1979a). These kilns generally appear to be square or rectangular, double-chambered, roofless structures, although one round brick kiln was recorded in southern Italy (di Caprio 1979:91). Sicilian tile industries are becoming increasingly scarce as mass-produced sheet roofing replaces traditional tiles (Carter 1979:62).

In modern Mexico, however, single-chambered, modified clamp-type kilns appear to be the universal choice. The brickmaking process has been described in general by Boudreau (1974) and Tays (1941) and more specifically for manufacturing sites in Guadalajara (Rice University 1966) and Tecate (May 1970). In these accounts, the mixing of clays and the fabrication of bricks are quite similar, and the resulting products are nearly identical to both the Roman examples and those from Mission San Antonio.

Spanish Kilns at Tejar, Guatemala

In modern Guatemala, the same Spanish colonial brickmaking technologies, including use of single-chambered kilns, also seems to have survived. San Miguel de Tejar is located in the highlands of Guatemala, about 55 km northwest of modern Guatemala City. The vicinity was early recognized by Spanish colonizers as an excellent source of clay. Tejar ("tile works") was established as an early tile-production center for the colonial-period capital of Antigua, founded in 1534.

In 1978 the tile and brickmaking component of the modern town of Tejar consisted of a scattered collection of 30 to 40 kilns each with its own clay quarry, clay-processing area, brickyard, and in most cases, the residence of the brick maker and his family. The kilns are of the same modified clamp-type recorded in Mexico and excavated at Mission San Antonio. The *tejas* and *ladrillos* produced are virtually identical to Roman, Mexican, and Alta California examples.

Clay deposits make up the subsoil of the settlement and clay pits are in evidence throughout the area, larger ones said to have been in use for hundreds of years. These pits range from house-size excavations to quarries as large as a quarter acre. The excavated clay is removed to the processing area where it is watered, kneaded, and stored in a large pile near the manufacturing area. No temper is added to the raw material.

Wooden molds are used to form the tiles. A single or three-*ladrillo* mold is used for the square tiles and a single, tapered rectangle for the *tejas*. *Ladrillo* molds are dipped in water to keep the clay from sticking and then placed on the ground in the open drying area. Clay is pressed into the mold, corners are packed and the surface hand smoothed before the mold is removed and placed in the open end of the row where the process is repeated. *Tejas* are molded on a table and then transferred to curved forms which have been moistened and coated with sand to facilitate their removal in the drying area. To be dry enough to handle, the tiles require two days in the summer and eight days in the winter. When firm, they are stacked aside to continue drying until ready for firing.

The kilns of Tejar are remarkably alike, only varying in size according to the wealth of the brick maker. Because of Tejar's location in an active earthquake area, kilns reportedly last only about 20 years. During the 1978 study, two family-operated kilns were inspected in detail and interviews conducted with the owner-operators.

Figure 2. The kiln of Juan José Quá, Tejar, Guatemala.

Juan José Quá lived next to his brickyard with his wife and three children. He worked his kiln year-round although the dry summer months were more productive. His kiln was two years old, rebuilt after the devastating earthquake of 1976. It was a rectangular (3.6 m x 2.75 m) modified clamp-type kiln: single-chambered (internal area of 4.6 m^2), with partly temporary walls (Figure 2). Permanent walls were made of mud-mortared adobe blocks and were .43 m thick and 2.15 m high. When the kiln was stacked for firing, a temporary wall extension of unmortared *ladrillos* was added to the structure bringing the total kiln height to 3m. The kiln was built into an old clay pit, the embankment of which served as the rear structural support as well as provided access to the upper part of the kiln chamber. Slot "windows" were built into the sides of the kiln to facilitate stacking and unstacking the load. Before firing they were closed with *ladrillos*.

Two arched firemouths (.86 m x .61 m) were symmetrically located on the downhill face of the kiln. For firing, green bricks were carefully stacked in the kiln chamber to promote even circulation of heat and gases, forming a long hollow passage, the central flue, inside each opening. Unlike the Romano-British modified clamp kilns, this flue is not a permanent trough in the floor but was formed by the stacked green wares. Firewood was laid along the length of this passage during the stacking process, extending just outside the firemouth. Above the flues, paired bricks with 5 cm spaces between them were stacked in a lattice fashion to the height of the kiln walls. All types of brick and tile could be included in the same firing. When loads were mixed, *tejas* and other more fragile forms were usually placed near the top.

The area in front of the firemouth was covered with a shed-like awning to keep the fuel and work area dry during the rainy season and provide shade for the brick maker during the hot summers. Available wood was *cipré* (cypress), *gravilaia* (?) and *pino* (pine), all of which was utilized. Quá noted, however, that *pino* leaves a white coating on the bricks and is therefore not as desirable. Once sufficient numbers of tiles have been prepared and thoroughly dried, they were fired. The kiln held approximately 6400 bricks and took about one day to stack. The windows were bricked up and a tin roof added to the top. Traditionally the roof was made of baked tiles but modern tin is easier to handle.

The burning took about 12 hours. Fuel was added according to the quantity of smoke exiting from the sides of the tin roof and the color of the bricks as checked from the side windows. When firing was complete, the kiln was left to cool for another day and then emptied. The "best color" (a bright red) was attained by the bricks in the center of the kiln which were sold as higher quality than those near the edges. Quá estimated that he had a loss of only 20 or 30 bricks per firing. These were accumulated in a waste heap near the kiln but out of the work area.

The kiln of Juis Ugho was similar in construction and use to that of Quá. It was larger, measuring 3.75 m on a side with an interior chamber of 5.5 m^2. The permanent adobe walls stood two meters high with a .5 m-high wall extension of baked brick erected for firing. Ugho's kiln had twin arched firemouths similar to Quá's and also had opposing stacking windows on the side walls.

Kiln walls were .75 m thick. A .6 m-square buttress was added to one corner following the 1976 earthquake. Earth had been piled on three sides of the free-standing kiln to a height of .6 m, likely for both wall stabilization and insulation. Kiln capacity was 20,000 bricks and firing time 24 hours. Ugho informed us that only three hours were needed for cooling although this seems to be quite a short period. He reported that "*grabadora*" was the preferred wood (this may be same species as the "*gravilaia*" wood mentioned by Quá).

The techniques of tile manufacture, as manifest on the finished products, do not appear to have changed from Roman to modern times. The single-chambered kilns at Tejar compare with the smallest examples of this type from Roman Britain, lacking only the recessed central flue. They are more similar to the example from Mission San Antonio, which also lacks the central flue and is only slightly larger. The processes observed at Tejar may, therefore, assist in reconstructing these activities in Spanish California.

TILE MAKING IN SPANISH CALIFORNIA

The major objective of the Spanish settlement of Alta California in the 18th century was to secure the land as part of the Spanish empire. The colonizers were to "civilize" the native populations who would then maintain and defend the Hispanic towns and economies from the increasing attentions of the British and Russians. To accomplish this, self-sufficient mission communities were established, run by Franciscan priests and policed by Spanish soldiers. Local native peoples were gathered into the settlements and taught the technological, social, and religious aspects of Spanish culture. Ideally, when the lessons had been properly learned, the padres would move on to other unconverted populations, leaving a self-sufficient town. Altogether, 21 missions and four presidios were established along the coast between 1769 and 1823. The Mexican revolution (1810-1821), however, effectively cut off the colony's political and financial support from Spain. The new Revolutionary government subsequently honored its mandate by secularizing the missions (between 1832-1834) and disbursing church lands to private individuals (Costello and Hornbeck 1989).

Whitewashed adobe buildings with roofs, floors, and corridors of red tile are architectural hallmarks of Spanish California, an integral part of its Hispanic architectural heritage. Like all other imported European technologies, local manufacture of fired tiles required skilled workers to train the Native Americans. For Alta California, virtually all these individuals came from long-colonized Mexico, brought by the Spanish government as either soldiers or civilians. *Ladrilleros* (or *tejeros;* tile makers) are not identified specifically in historic documents (Schuetz-Miller 1994), and evidently did not rank as highly as carpenters, blacksmiths, masons, and other artisans of the building trades. The first tile maker in Alta California evidently first worked at Mission San Antonio, where fired tiles are first referenced (Engelhardt 1972:22). Soon the other California missions began recording the use of tiles at their establishments. This individual soon moved on to other institutions and other tile makers were also likely recruited for short periods to assist in introducing the technology. After an initial training period, the missions were apparently left on their own to carry on the industry.

For several reasons, each Alta California mission developed its own tile works. The acceptability of commonly occurring residual clays, tempered where necessary, permitted production to be set up in virtually any location along the coast (Newcomb 1925:75). The simple, labor-intensive process made the technology appropriate to the missions where there were substantial numbers of unskilled personnel. It is also unlikely that the large quantities of tiles utilized at each mission were transported any great distance. A reliance on solid-wheel carts and roads generally designed for foot and hoof traffic would have dictated local manufacture of the heavy, bulky, and breakable *tejas* and *ladrillos.*

Historic Spanish documents rarely address the mundane process of tile making. Letters and annual reports from the mission Fathers dealt primarily with topics such as the spiritual state of the neophytes, the problems of supply shipments from Mexico, and the exploitive activities of the local military population toward the native converts. This latter theme prompted an account of Indian labors, written by the Fathers of Santa Barbara Mission in 1800, in response to accusations of poor management by the Comandante of the Santa Barbara Presidio:

> "...Those who make tiles also have their duties assigned. Sixteen young men are selected, and sometimes as many more middle-aged men, or recruits from the *rancheria* [Indian village]; all these people, with two women who carry them sand and cow dung, make 500 tiles a day. They

> have the troughs of clay nearby, and these are always full. These Indians finish working before eleven; ... In 1795 the Comandante of the Presidio, Don Felipe Goycoechea, asked for ten Indians to make tile. .... After laboring four days at the Presidio, on Thursday they complained that they could not endure the work, and that their hands and arms were very sore. They were asked how much work they did per day. They replied, 500 tiles; which meant that they had to dig the clay and throw it in the troughs; that they had to add cow dung, and water from a well 15 varas [12.6 m] deep, or from a lake some distance away; that they had to carry the lake water from the beach, an eighth of a league [.5 km] from where they made the tile; that they had to mold the clay; and finally, that they alone, between morning and night, had to produce 500 well-made tiles, for the soldier Olivas stood over them to see that the work was done" [Egenhoff 1952:159].

Additional tasks in the tile making process were also addressed in this same account of labors:

> "All of the women who are considered useful help carry adobes...These same women help carry the brick and tile, and very infrequently stones—but only small ones—for leveling foundations...Of the little children over nine years of age, some comb the wool in looms and hand the shuttle and quill to the weavers; others look after the tile and brick during the day, so the animals do not step on it" [Egenhoff 1952:159].

That Franciscan fathers themselves may have been required to serve as *ladrilleros* is evidenced in this description by the French visitor Duflot de Mofras in 1841:

> "One day when I went to see him (Fr. Tómas Esténaga, San Gabriel Mission), I found him in a field before a great table, his hood thrown aside, his sleeves pushed up, molding clay in his hands and demonstrating to the neophytes that surrounded him, how to make bricks" [Egenhoff 1952:185].

Some manufacturing procedures can be generalized from this account: local soils were used and these were tempered with sand and organic matter; the clay was wedged in earthen troughs, and the tiles were molded and laid outside to dry in a manner similar to that in Tejar. An undated inventory taken at San Francisco de Solano Mission lists *"...2 ladrilleras de madera"* and *"8 moldes de teja"* ("...2 *ladrillo* molds of wood" and "8 molds for roof tiles") (Webb 1952:109). While the padre described above by de Mofras is certainly working his clay into a mold, evidence of manufacturing techniques is also contained on the historic tiles themselves: finger impressions from packing and smoothing clay, sand on the undersides of *tejas* to free them from the curved mold, and animal tracks left on soft *ladrillos* in the drying area.

Tile sizes in Spanish California were based on increments of a *vara*, which at that time equaled 83.8 cm. Archaeological examples of *tejas* reveal lengths ranging between 50-60 cm; the larger curved end measuring from 20-33 cm long, bending upward some 10-15 cm, while the smaller end measures 15-23 cm and the curve is 7-10 cm in height. Roof tiles are 2-4.5 cm thick. *Ladrillos* are generally square, the most common is 27 cm on a side (1/3 *vara*), although they range between 15x15 cm and 30x30 cm. Rectangular tiles were formed by cutting any square tile in half, or by using a rectangular form based on a ratio of 3:2 with the long side measuring between 31-35 cm. Triangular tiles were formed by cutting a square tile diagonally. More elaborate shapes, such as those recovered from the second site of La Purísima Mission (Hageman 1980:127) identify the hand of a *ladrillero* from New Spain. *Ladrillo* thickness ranges from 3.8-5.7 cm, but 4-5 cm is most common. Size variation was common not only between but within mission institutions as tile molds were not standardized and clays from different locations had different shrinkage rates.

In spite of the widespread use of fired tiles and their production at each California mission, only two kiln sites are presently known. The brick-and-tile kiln at Mission San Antonio, built in 1780, was excavated in 1976-77 as part of a field school (Costello 1985). The kiln site at Mission La Purísima Vieja in Lompoc has been located but not investigated (Costello 1993). Other kiln sites are only known through historical reference. In the first decades of the 20th century, John P. Harrington noted that "...a kiln above Santa Ynez was used only for making tile and bricks" (Browne 1974:63); this facility has never been located. At San Juan Capistrano "...the tiles and bricks for the construction of walls, arches, and pavements, as well as the roofing-tiles, were made just north of the mission, where remnants of the kiln may be seen to this day" (Newcomb 1925:155). This kiln in *la Cañada del Orno* (sic) (Kiln Canyon) was reportedly destroyed in ca. 1925 (Webb 1952:110). Two kilns stood at Santa Barbara Mission, one for firing tiles and one for firing pottery (Webb 1952:120), and a tile kiln is evident in an 1870 photo of Mission San Buenaventura (Browne 1974:Plate 3). Santa Barbara and San Buenaventura, however, have been subjected to steady urban development and the remains of these kilns seem to have been destroyed.

Mission San Antonio

Mission San Antonio de Pádua was founded in 1771 in the San Antonio River Valley in central western California. At the time of its founding, there were estimated to be close to 3000 Salinan Indians occupying the region between Monterey and San Luis Obispo. Salinan economy was based on the collection of vegetal foods, especially acorns, with shellfish, deer, rabbits, and fish also comprising important parts of the diet. The industries of the Salinan included basket making, stone working (ground and pecked stone as well as flaked), and the manufacture of wooden tools and implements. The bow and arrow was used in hunting, clothing was of woven plant fiber and fur, and houses consisted of a domed pole framework covered with tule or rye grasses. There was a shell-based currency and the Salinan had both calendrical and numerical systems, and methods for linear measure (Hester 1978:500-504). Salinan culture did not include formal agriculture, animal husbandry, or ceramic production.

By 1790 San Antonio was the largest of the California missions with a population of over 1000 (Engelhardt 1972). By the first decade of the 19th century, the Mission was an extensive and diversified institution covering some 1200 square miles of land and including at least four outlying ranch stations (Costello 1994). The Mission population, industries, and stores were clustered around a plaza, at the head of which was the secure adobe quadrangle dominated by the church and containing workshops, storage areas, the residences of the Fathers, and a dormitory for unmarried women. Residences of the soldiers and Indians flanked opposite sides of the plaza along with agricultural and industrial work areas (Figure 3).

The ultimate fate of Mission San Antonio, along with all the missions in California, was decided by a proclamation of the new Mexican government in 1831, which effectively dissolved the Catholic Church's vast holdings of land and livestock. In 1834, the administration of Mission San Antonio was officially turned over to a governmental appointee, its extensive agricultural and livestock resources were appropriated by private parties, and most of the remnant Indian population drifted off into the countryside. In 1845, an inventory of San Antonio included some structures, a vineyard, an orchard, a garden, and four horses (Engelhardt 1972:79).

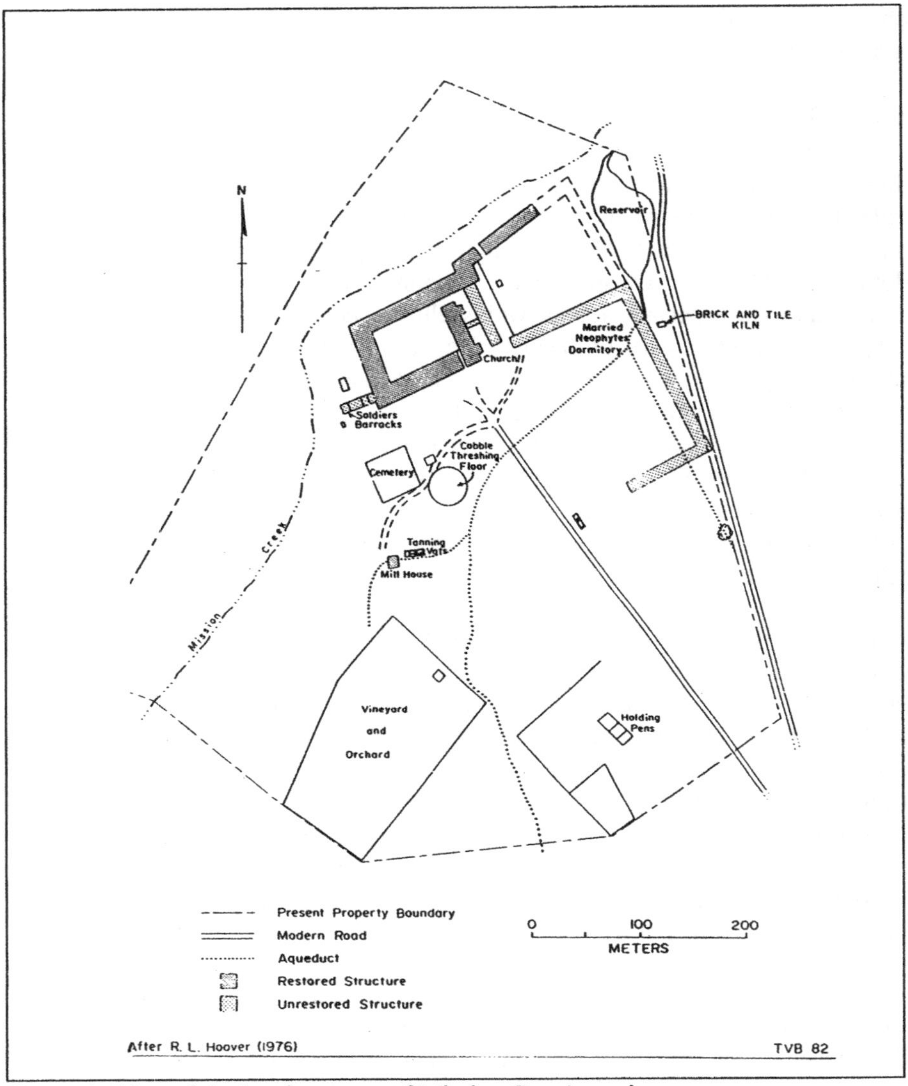

Figure 3. Plan of the headquarters of Mission San Antonio.

After the American Flag was raised in Monterey in 1846, the ownership of the Mission churches was conferred back to the Catholic Church. Until almost the turn of the century, San Antonio had a resident priest who maintained a small congregation and a few deteriorating buildings. Restoration of the church and quadrangle was completed in 1948-1949. Mission San Antonio is presently operated by the Franciscan Order and serves as both a parish church and retreat center.

THE KILN AT MISSION SAN ANTONIO

The location of the Mission San Antonio kiln site was easily identified by the large mound of tile wasters which had accumulated outside the kiln mouth. Archaeological excavations were directed toward uncovering the kiln structure and sampling the contents of the waster mound (Costello 1985).

The kiln is rectangular and is oriented at right angles to the hillside (Figure 4). It measures 4.0 m long and 2.1 m wide in the interior, tapering slightly at the firemouth; the highest preserved height of the adobe-block walls is 1.7 m. The internal chamber is 8.4 m^2 and is lined with alternating vertical columns of adobe blocks and stacked *ladrillos* (Figure 5). The absence of interlocking bonds between these columns may have allowed for expansion and contraction of the firing chamber during use. The interior surface of the chamber exhibited heavy vitrification and a glass-like veneer to a depth of approximately 2 cm. Heat from firing had oxidized the floor of the kiln to a depth of ca. 30 cm. The concave form of the floor—the center lying some 20 cm below the sides—may have served as a central flue, or may simply be a result of periodic cleaning.

The waster mound surrounding the kiln was sampled by excavating a trench 1.0 m wide and 3.5 m long into its front slope. Among the ash and charcoal, over 600,000 artifacts were recovered, nearly 500 kilos. Ninety-six percent of the artifacts were fragments of tile: 1/3 was pieces of *ladrillos,* 1/3 pieces of *tejas*, and 1/6 the remains of fired adobe bricks. The remaining tile pieces were too small or deformed for type identification, or were vitrified into globules of glasseous matter.

These tile fragments represent products rejected from the load for three primary reasons: overfiring, underfiring, or breakage. "Overfired" tiles showed signs of vitrification, indicating exposure to excessive kiln heat; they were typically brittle and bloated. "Underfired" tile fragments were occasionally recovered but

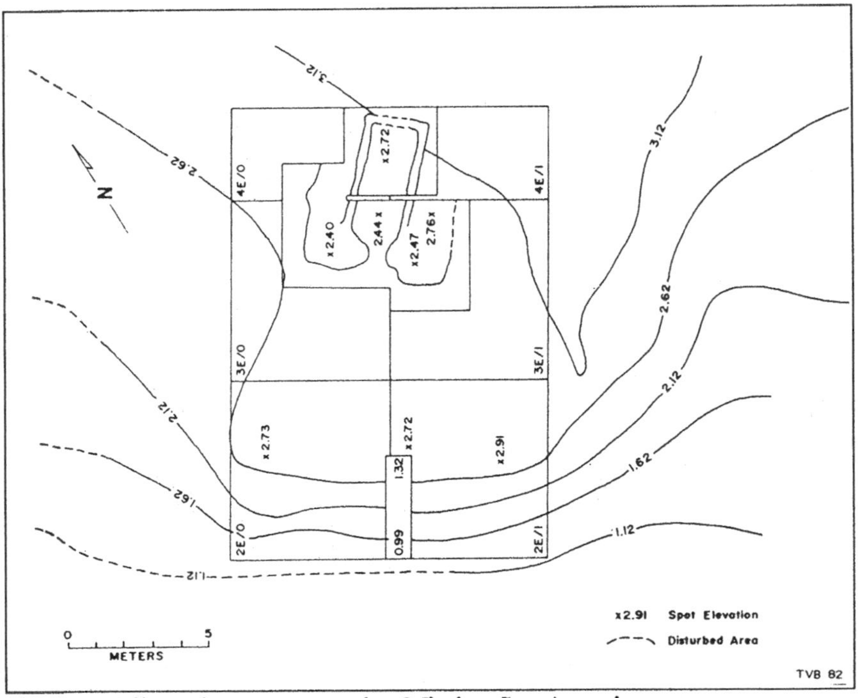

Figure 4. Kiln and waster mound at Mission San Antonio.

are numerically insignificant as they tend to break down quickly into the soils; whole, underfired tiles may also have been restacked with the next kiln load. Both overfiring and underfiring result from a lack of heat control during the firing process, both in maintaining a constant temperature and in the distributing heat evenly around the firing chamber. Adequately fired waster-tile fragments were classified as "broken," indicating that they were not discarded due to their firing state but presumably because they did not emerge from the kiln intact. Breakage could be due to mishandling of the wares, or to large inclusions in the clay which weaken the structure. Pieces of completely vitrified, glasseous material are clay fragments melted from repeated exposure to firing temperatures, a process also seen in the vitrified surface of the kiln walls. As these metamorphosed tile pieces must have remained in the kiln for successive firings, the chamber was apparently not thoroughly cleaned between loads. The fired adobe bricks were recognized as the kiln's temporary roofing material, thrown away when they became oxidized and unserviceable.

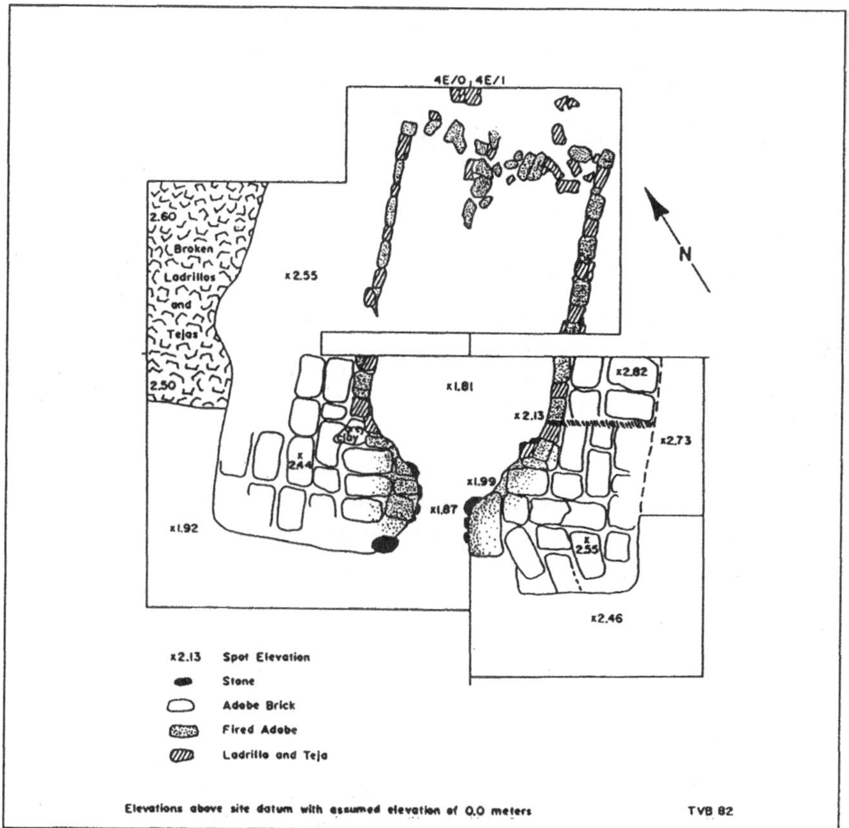

Figure 5. Plan of kiln at Mission San Antonio.

Tile Color and Kiln Firing Temperatures

To determine the actual temperature ranges present in the historic kiln, small test bricks were manufactured from local clays and fired in an electric kiln at regularly increasing temperature increments (Costello 1985:Table 66). Color was used as a guide to evaluating the kiln products:

> Where bricks are made from a shale or clay that is consistent in composition and nature, the color of the burned brick provides a very useful guide to the degree of burning and, therefore, to the water absorption, strength and durability of the bricks. Red burning (i.e. iron-bearing) clays almost universally have a depth of color proportional to

> the degree of burning. The darker colors are usually associated with the harder or burned brick, and the "salmon" color with the underburned brick. ...Color alone does not suffice for a grading of different makes of brick, but is quite valuable for the purpose of grading a shipment or kiln of the product of one manufacturer [Plummer and Rearden 1943:46].

The primary causes of pottery color are the composition of the clay, the atmosphere of the kiln, the duration of firing, and the temperature (Shepard 1971:103). These variables were controlled as much as possible during the test firings to enhance comparability with historic kiln conditions. Temperature was measured by pyrometric cones as a preferred thermoelectric pyrometer was not available. Each firing arbitrarily spanned several hours in an effort to duplicate the presumed longer heating rate of the historic kiln.

Fired clay is principally colored by the amount, particle size, and distribution of iron oxides, once any organics have been burned out (Shepard 1971:103). As the soils around the kiln site were certainly the source of the historic clays, these were used for the experiment. Sifted to remove large inclusions, the clay was formed into small loaves (15 x 6 cm) with a thickness approximating that of the historic tiles (2.5 cm). Dried in the shade, they could be handled in two days and were dry enough for firing in five.

The amount of oxygen in the atmosphere of the kiln during firing is the second major variable that affects ware color. When sectioned, historic tiles in Alta California typically reveal layers of colors surrounding a gray or black core. When pieces of historic tiles were refired at temperatures of 700°C the black cores disappeared, confirming that the original firings were insufficient in temperature, time, and draft to burn out the organic matter in the clay.

The absolute temperature at which a clay is fired to a "mature" (optimal) state varies according to mineral constituents and kiln atmosphere. Within any one clay, however, the mature state is indicated by the clarity of the color when fired in a fully oxidizing atmosphere (Shepard 1971:214). This clarity can be determined from the Munsell color scale for "chroma," which will attain its highest value when a fired clay is mature and will decrease again with overfiring. The results of the test-brick firing indicate that the optimum temperatures at which the Mission tile-clay matures in an oxidizing atmosphere is between 1040°C and 1125°C.

The range of colors present in historic tiles from mission structures were compared to this optimal range. It appears that the mission builders had a generous definition of a serviceable tile, which ranged from being fired in temperatures as low as 660°C, below which the sample could be dissolved in water, to as high as 1170°C, where there was initial sintering and fluxing. In kilns where firings are not well controlled, all of these temperature ranges may be evident within a single lode.

CONCLUSIONS

The Spanish New World brick-and-tile kilns differ from their Roman antecedent in being somewhat smaller and in single-chambered kilns being the predominant kiln type. Single-chambered kilns are not unknown on sites belonging to the Roman Empire, but are unusual, and are accompanied by a recessed and permanent central flue, a feature apparently lacking in *Nueva España*. The kilns at Mission San Antonio and at Tejar were similar in size and function, those at both locations being single chambered, wood fueled, and covered with temporary roofing material. Details of kiln operation provided by the Guatemalan *ladrilleros* likely reflect procedures also relevant at the historic mission. The process of forming *tejas* and *ladrillos* appears similar in all locations and time periods, documented by the physical attributes of the archaeological examples.

Study of the current tile making operations at Tejar has also broadened our understanding of the types of work spaces required by California mission brick and tile industries. In addition to the kiln and clay source, areas were needed for cleaning, pugging, and storing the clays. Large flat areas were necessary for laying out and drying the greenware; in hot weather, protective shelters may have been needed for shade. As most mission kilns were sited immediately adjacent to the main settlement (and therefore close to construction activities), space could soon become a problem as the mission complex grew. At Mission San Antonio, the locations of the various activities related to tile making changed as the Salinan residence area was eventually extended to the foot of the kiln waster mound.

When compared with its Guatemalan counterpart, the technology of brickmaking at the California mission appears inexpert. A contributing variable in this difference is the substantially inferior clays utilized by the mission tile makers. While Tejar's brickworks were purposefully established in the middle of excellent clay deposits, the Alta California kilns were part of the mission headquarters, and these were sited for proximity to large Indian populations. The brickmakers at the missions had to make due with what was locally available, the best of which was frequently the clays and silts which accumulated in the mission reservoirs. Many of the identified kiln sites are adjacent to these facilities, which served as great settling ponds and provided perhaps adequate but not excellent raw material.

Skills exhibited in kiln firing also appear to be much more proficient in Tejar than at San Antonio Mission. While waste material at Tejar was minimal, a mere 50 years of kiln operation at San Antonio had produced a substantial mound of rejected tiles in front of the kiln mouth. This large discard pile accumulated despite generous levels of acceptability by the builders at Mission San Antonio, who utilized tiles ranging in fired states from barely oxidized to near vitrification. Evidence for poor control of kiln heat during firing is also found in the tiles themselves: while those of Tejar were homogeneous in section, Alta California tiles exhibit variations in colors that document poor control of both temperature and heat circulation within the kiln.

Perhaps the most important reason for this apparent contrast in expertise is the short duration of the California colony. The skilled brick and tile makers imported from New Spain stayed only long enough to establish a kiln at each mission and teach basic techniques to their apprentices. Each mission then had to develop its own experts. While Tejar had from 200-400 years for indigenous people to adopt Spanish culture and for local craftsmen to become proficient in the introduced technologies, Alta California had about only 50 years.

The colonization effort in this last Spanish frontier was cut off before Hispanic institutions had been fully adopted and mastered by resident populations. One measure of this superficial conversion was the wholesale abandonment of the construction tile industry once the yoke of mission life was lifted from the native peoples. After the dissolution of the missions ca.1835, new adobe constructions and repairs to older buildings utilized scavenged tiles from abandoned mission buildings. The few trained *ladrilleros* (Mexican and Indian) evidently had other preferred occupations, or demand was low, or rewards meager. Spanish culture was apparently not imposed long enough in this colony to produce a self-maintaining Hispanic community. Ready abandonment by native peoples of the red-tiled roofs and deep brick corridors of Spanish and Mexican adobes is one measure of this.

REFERENCES

Blagg, T.

1979 The Use of Terra-cotta for Architectural Ornament in Italy and the Western Provinces. In *Roman Brick and Tile*, edited by A. McWhirr, pp. 267-284. B.A.R. International Series 68. Oxford, England.

Boudreau, E.

1974 *Ways of the Sierra Madre*. Pleasant Hill Press, Sebastopol, California.

Browne, R. O.

1974 San Buenaventura Mission Water System. Report submitted to the Ventura City Council in partial fulfillment of an agreement with the Ventura County Archaeological Society, June.

Carter, J.

1979 Rural Architecture and Ceramic Industry at Metaponto, Italy, 350-50 B.C. In *Roman Brick and Tile*, edited by A. McWhirr, pp. 45-64. B.A.R. International Series 68. Oxford, England.

Costello, J. G.

1985 The Brick and Tile Kiln, in *Excavations at Mission San Antonio, 1976-1978*, edited by R. L. Hoover and J. G. Costello; pp. 122-145. Monograph 26, Institute of Archaeology. University of California, Los Angeles.

1993 *Mission Vieja de La Purisima, CA-SBA-521H, Report on the 1991-1992 Archaeological Investigations.* Community Development Department, City of Lompoc, Lompoc, California.

1994 *The Ranches and Ranchos of Mission San Antonio de Padua.* California Mission Studies Association, Keepsake Volume, December 1994.

Costello, J. G., and D. Hornbeck.

1989 Alta California: An Overview. In *Columbian Consequences, Volume I: Archaeological and Historical Perspectives on the Spanish Borderlands West,* edited by D. H. Thomas, pp. 303-331. Smithsonian Institution, Washington, DC.

Cram, L., and M. Fulford

1979 Silchester Tile Making—The Faunal Environment. In *Roman Brick and Tile*, edited by A. McWhirr, pp. 201-210. B.A.R. International Series 68. Oxford, England.

di Caprio, N. C.

1979 Pottery- and Tile-Kilns in South Italy and Sicily. In *Roman Brick and Tile*, edited by A. McWhirr, pp. 73-96. B.A.R. International Series 68. Oxford, England.

Dymond, D. P.
1974 *Archaeology and History: A Plea for Reconciliation*. Trans-Atlantic Arts, Levittown, New York; and Thomas and Hudson, London.

Engelhardt, Fr. Z., O.F.M.
1972 *San Antonio de Padua*. Ballena Press, Ramona, California.

Egenhoff, E. L. (ed.)
1952 *Fabricas*. Supplement to the California Journal of Mines and Geology, California State Department of Natural Resources, Division of Mines, Sacramento, California.

Hageman, F. C.
1980 An Architectural Study of the Mission La Purísima Concepcion, California. In *An Archaeological and Restoration Study of Mission La Purísima Concepcion*, F. C. Hageman and R. C. Ewing, pp. xxii-234. Santa Barbara Trust for Historic Preservation, Santa Barbara, California.

Hester, T. R.
1978 Salinan. In *California*, edited by R. F. Heizer, pp. 500-504. Handbook of North American Indians, vol. 8, W.C. Sturtevant, general editor. Smithsonian Institution, Washington, DC.

Lister, F. C., and R. H. Lister
1977 *A Descriptive Dictionary for 500 years of Spanish-Tradition Ceramics (13th Through 18th Centuries)*. Special Publication Series 1. Society for Historical Archaeology.

May, R. V.
1970 Letter describing tile making in Tecate, Mexico; with slides. In possession of the author.

McWhirr, A.
1979a *Roman Brick and Tile: Studies in Manufacture Distribution and Use in the Western Empire*, edited by A. McWhirr. B.A.R. International Series 68. Oxford, England.
1979b Tile-Kilns in Roman Britain. In *Roman Brick and Tile*, edited by A. McWhirr, pp. 97-190. B.A.R. International Series 68. Oxford, England.

Newcomb, R.
1925 *The Old Mission Churches and Historic Houses of California*. J.P. Lippincott, Philadelphia.

Peacock, D.
1979 An Ethnoarchaeological Approach to the Study of Roman Bricks and Tile. In *Roman Brick and Tile*, edited by A. McWhirr, pp. 5-10. B.A.R. International Series 68. Oxford, England.

Plummer, H. C., and L. J. Reardon
1943 *Principles of Brick Engineering*. Structural Clay Products Institute, Washington, DC.
Rice University
1966 *Vaulted Brick Construction in Guadalajara*. Architecture at Rice University, Vol.18. Houston, Texas.
Santa Barbara Mission
1834 Inventory.
Schiffer, M. B.
1978 Methodological Issues in Ethnoarchaeology. In *Explorations in Ethnoarchaeology*, edited by R. A. Gould, PP. 229-247. University of New Mexico Press, Albuquerque.
Schuetz-Miller, M. K.
1994 *Building and Builders in Hispanic California*. Southwest Mission Research Center, Tucson, and Santa Barbara Trust for Historic Preservation, Santa Barbara, California.
Shepard, A. O.
1971 *Ceramics for the Archaeologist*. Publication No. 609, Carnegie Institution, Washington, DC.
Sölter, W.
1970 *Romische Kalkbrenner im Rheinland*. Rheinland-Verlag, Düsseldorf.
Tays, G.
1941 *Ranch and Mission Industries in Spanish California*. Unpublished Ph.D. dissertation, University of California, Berkeley, California.
Webb, E. B.
1952 *Indian Life at the Old Missions*. Warren F. Lewis, Los Angeles.
Webster, G.
1979 Tiles as a Structural Component in Buildings. In *Roman Brick and Tile*, edited by A. McWhirr, pp. 285-294. B.A.R. International Series 68. Oxford, England.
Wilson, R.
1979 Brick and Tiles in Roman Sicily. In *Roman Brick and Tile*, edited by A. McWhirr, pp. 11-44. B.A.R. International Series 68. Oxford, England.

AN INTEGRATIVE REVIEW AND EXAMINATION OF GLASS FURNACE TECHNOLOGY IN RENAISSANCE ITALY

W. Patrick McCray
Program on Culture, Science, Technology, and Society
College of Engineering
University of Arizona
Tucson, Arizona 85721

ABSTRACT

Several different sources of evidence exist which can be used to make inferences regarding Renaissance glass furnace technology. These include literary citations, pictorial representations, and material analyses. Information from the aforementioned sources is supplemented with evidence drawn from the author's chemical and physical analyses of Renaissance Venetian glass vessels and sherds. In this paper, the various lines of evidence are drawn together to provide a more comprehensive look at the science and technology of glass furnaces as constructed and used in Renaissance Venice.

INTRODUCTION

It is a generally accepted principle that with the fall of the Roman Empire the glass industries of northern and southern Europe began to develop along separate technological paths (Polak 1975:37-50; Tait 1991:110-111). This geographic variability is observed in several ways including the use of different raw materials, the preferred location of glassmaking industries, and the types of objects manufactured. Another distinction that emerged was the varying manners in which glass furnaces were constructed in northern Europe versus those in southern Europe and the Mediterranean region. The prevailing distinction in glass history studies has been that glassmaking in northern Europe was carried out in primarily rural locations using "rectangular" furnace designs while glass production in the south was mostly concentrated in urban areas using a "beehive" form of furnace construction.

This study presents a review of material relevant to the design of glass furnaces constructed in the "southern" tradition. Particular attention is paid to the glass industry of Renaissance Venice. The craft of glassmaking in Venice has been concentrated in the glass factories of Murano since at least the 10th century. This island community was arguably the most important locale for glassmaking between the 15th and 17th centuries. In addition to design considerations, inferences regarding the nature of glassmaking crucibles and staffing and operation of Renaissance-era glass furnaces are presented. In this manner, a more comprehensive picture of glassmaking technology in both its social and technical context emerges.

THE DESIGN AND CONSTRUCTION OF RENAISSANCE GLASS FURNACES

The variety of available sources pertaining to Renaissance Italian glass furnace design and operation stands in contrast to some of the other studies presented in this symposium which have had to rely primarily on the interpretation of archaeological material. This is a fortunate situation for one interested in past technologies. This evidence includes documentary sources (technological treatises, recipe books, and travelers' records), pictorial representations of glass furnaces, and the analysis of glass and refractory materials.

In comparison with other aspects of glassmaking technology in Renaissance Venice, a fair amount of information concerning furnace design has been presented previously (Charleston 1978; Newton and Davison 1989 for example). Much of this is description based on surviving documents, pictures, and some archaeological work. A review and integration of this material into the larger context of production is given here. The previous work, however, does neglect certain other issues of furnace technology. One is the question of the evolution of furnace design and its accompanying variability. Another is the specialized nature of the furnaces, their operation, and the production and use of glass crucibles.

A very early indication of the basic shape Venetian glass furnaces would eventually assume is found in a Syrian manuscript held at the British Museum. Dated no earlier than the 9th century A.D., it describes a furnace round or cylindrical in shape and consisting of three levels. The bottom was a fire chamber, the middle was the central story in which the glass was melted and gathered, and at the top there was a third compartment in which the finished vessels were annealed (Charleston 1978:11). The earliest known illustration of a glass furnace in the southern tradition, found in the *De Universo* manuscript held at the Abbey of Monte Cassino, confirms the impression given by the Syrian document. What appears to be a cylindrical structure, based on the elliptical glory holes (where the glass was gathered and worked) in the walls of the furnace, is depicted with three different levels. This is shown in Figure 1.

Figure 1 - Earliest known illustration of a "southern" style glass furnace. This is taken from the *De Universo* manuscript (c. 1023 A.D.); it shows a three level furnace in the style that would become common in Renaissance Venice.

An early site in northern Italy that has yielded evidence of glass or evidence of glassmaking is Torcello located in the Venetian lagoon (about 8 km from the modern-day city of Venice). The site was excavated in 1961-62 by a Polish-Italian team and the entire project is described in a 1977 monograph (Tabaczynska 1977). Near the center of the small island, adjacent to the present day Basilica of Santa Maria Assunta, the excavators claim to have uncovered the remains of four early medieval furnaces, along with tools and waste associated with glass production.

Four different furnace structures were unearthed including one of circular form about 3.2-3.5 m in diameter. One scholar has concluded that there was a continuity in furnace design and methods of production between the Roman era and the high Middle Ages (Gasparetto 1967:75). The first furnace was for the preparation of the glass frit, the second and third were for the melting and working of the glass, while the fourth was for annealing (Tabaczynska 1968). The glass furnaces have been associated with the production of mosaics for the nearby basilica which dates to 639 A.D. The excavators have suggested that the furnaces were destroyed or transferred elsewhere sometime in the 7th century.

Excavations carried out in northern Italy at later sites provide additional evidence for the design and construction of Renaissance-era glass furnaces. For example, a furnace excavated at Monte Lecco near Genoa was dated to the late 14th or early 15th century (Mannoni and Fossati 1975). Service and industrial areas were identified including the remains of an "igloo-shaped" furnace which had a variety of glass waste associated with it. A circular fire trench ran between two solid furnace sieges on which the glass-pots would have been placed. As a result, a three-storied structure like those described above seems most plausible. This was not an urban glasshouse as Venice's were but rather it was located in a forested mountain environment. Presumably, the nature of production activities as well as demand for glass would have been different due its more remote location. Gradually, the forest glasshouses of this Italian region were abandoned and glass production in this region became a more urban-oriented activity in the later 15th and 16th centuries (Calegari and Moreno 1975). Another excavation has revealed the presence of a glass furnace dating to the latter half of the 16th century at Pisa (Redi 1991). One of the buildings uncovered was a circular furnace structure with a suggested diameter of about 3 m. A relatively recent collection of papers contains several additional references to archaeological excavations connected with Italian pre-industrial glassmaking including the discovery of furnace remains in Tuscan and Mantuan locales (Mendera 1991).

An explicit understanding, based on archaeological excavations, of how glass furnaces were designed in Renaissance Venice is unfortunately lacking. There have been no published excavations in and around Venice (other than Torcello) which have yielded evidence of past furnace structures. In large part, this is due to the fact that archaeology in Venice is still not well established. Moreover, glass is still made at Murano, quite possibly in the same locations where glass production took place in the Renaissance. Past furnaces have long since been destroyed and new buildings erected in their place making excavations difficult if not impossible.

Documentary sources coupled with iconographic evidence from the Vatican Library gives a very detailed picture of how Italian glass furnaces were built and what they looked like in the late 15th and early 16th century. Two illustrations from late 15th-century Vatican manuscripts show glassblowers sitting on stools in front of the furnace. A stoking hole is at the bottom level while the two upper stories are depicted as being slightly set back from the first. In this manner, a ledge running around the circumference of the furnace was created on which tools and the marvering slab could be placed. In the early 16th century, a Swedish priest, Peder Månsson, compiled an account of the contemporary Roman glass industry. While different in some respects from glassmaking as practiced in Venice, his description of the Roman furnace is similar enough to written descriptions of Venetian structures to be considered here (Månsson 1520).

Månsson mentions two furnaces in his travel diary. One was the working furnace in which the glass was melted and gathered. The other, presumably, was the fritting furnace where the ash and sand were first heated together. This was the practice of preparing the raw materials in Venice. The primary furnace where the glass was melted and worked is described as circular and about 3.5 m in diameter. This is comparable in size to those at Torcello, Monte Lecco, and Pisa. The furnace, as depicted in the Vatican drawings, has three levels - one for the fire, one for the glass pots, and an annealing chamber where the finished products were cooled.

In 1540, Biringuccio published his treatise *De la Pirotechnica* (printed in Venice, incidentally). Along with describing the process of glassmaking, the author also discusses how the furnaces used in Venice were designed and assembled. The first furnace (known as the *calcara*) was used for making the glass frit is described as (Smith and Gnudi 1942:128):

> ... a reverberatory furnace made for this purpose, three *braccia* long (a *braccia* was about 55 cm), two wide, and one high, and apply enough of the strong flames of a wood fire by means of a reverberator so that this composition is melted well and converted all into one mass. When this operation is done it is cooled and then taken out and broken into pieces. This is the material that is called frit by the workers...

This type of furnace is shown in Figure 2, taken from Agricola's 1556 *De Re Metallica*. Note that there are only two levels to the *calcara*. The worker is shown in the process of breaking up the solid mass of frit.The second furnace (the *fornasa*) in which the primary melting and working is done is described by Biringuccio in a manner that coincides with the evidence presented thus far from the other sources:

> Now in order to complete the purification, a round furnace is made, built of rough bricks made from a clay that does not melt or calcine with fire.

Figure 2 - Fritting furnace as shown in Agricola's *De Re Metallica* (1556).

Jacoby, in his discussion of the raw materials used in the Venetian glass industry, notes that a Milanese petition of 1394 records the presence of "large slabs of dead stone from Custoza in the territory of Vicenza". At this time, Vicenza was under Milanese rule. It was under Venetian control by 1404 (Jacoby 1993:79-80). These large slabs of stone had a matte (dead) appearance and were supposedly used in the construction of various structures, including glass furnaces. While no examples of this material have been excavated and identified so that analyses can be done, it is imagined that these slabs of stone were desired for their refractory properties. Jacoby states that these slabs were used to make the bench in the main furnace which separated the heating and melting chambers (1993:79). Two Muranese documents of 1415 refer to these stone slabs (Zecchin 1990:181).

The round primary furnace described by Biringuccio had (Smith and Gnudi

1942:128):

> ... a vault with a diameter of four *braccia* and a height of six *braccia*. It is arranged in this way. First a passage for the fire is made which leads the flames into the middle of the furnace; around the circle at the bottom a shelf 3/4 *braccio* wide is made on which are to be placed the pots which hold the glass, and this must be one *braccio* above the ground. Around this five or six well-made little arches are built as supports for the vault, and under these are made the little openings which allows one to look inside and to take the glass out for working at will. Then the vault is continued to cover the glass...above this vault another vault is made which seals up and covers the whole...this is the cooling chamber...at the back of this is a trumpet shaped opening made into the circular shelf on top of the vault inside. The finished objects are placed here and can be skillfully drawn out with a long iron tool...after they have cooled.

Figure 3 shows a depiction of the furnace described by Biringuccio, again taken from Agricola's treatise. The "trumpet shaped opening" mentioned by Biringuccio is thought to be the beginnings of another element of the Italian furnace that would develop later - the *lehr* (Charleston 1978:17-18). This was the long, horizontal, and tunnel-shaped appendage to the glass furnace in which objects were placed to cool after the first cooling in the third story of the glass furnace. The objects were moved along this tunnel towards the opening, cooling ever more in the process. In a sense, the *lehr* is a derivation of the third story of the Venetian style furnace. By the 17th century, translations to Antonio Neri's *L'Arte Vetraria* were being published with illustrations showing a three-level furnace with a tunnel *lehr* extending from the back or side. Figure 4 gives an example, taken from an 18th-century French translation of Neri's book.

In the Palazzo Vecchio in Florence, there is a painting which is part of decorative cycle executed in the 1570s by Giorgio Vasari and his school (most likely by Giovan Maria Butteri). The concept behind the painting is the Aristotelian conception of air, earth, water, and fire. The paintings were used to decorate the studio of Francesco d'Medici. The Medici family had actively engaged Venetian glassmakers to come to Florence for the purpose of establishing glass factories as early as the late 16th-century (Heikamp 1986). On the side of the room corresponding to fire, the artist has represented some glassblowers laboring in front of a Venetian-style furnace (Barovier 1980:xliv). In addition to showing the manner in which 16th-century glassmakers carried out their activities, one may see the three-storied furnace as described in the above sources. The glass masters are shown

Figure 3 - Main melting and working furnace in "the southern" fashion as shown in Agricola's *De Re Metallica* (1556).

Figure 4 - Illustration showing the tunnel lehr as depicted in a 1752 French version of Neri's *L'Arte Vetraria*.

seated on three-legged stools before the furnace's glory holes with wooden slats strapped to their legs. These pieces of wood were the forerunners of the bench at which the glass master would sit. The pieces of wood provided support for the blowpipe and an easy place for the pipe to be rolled back and forth while it was rotated. Clay screens protect the workers. To the extreme left, one may see an assistant placing a finished piece in the top story of the furnace to cool. Workers in the forefront grind the raw materials and bring wood to the furnace. Toward the rear of the furnace is a tunnel-shaped appendage or overhang which may have been the precursor to the tunnel-*lehr* described above (Charleston 1978:16).

In short, a summary of the Venetian-style glass furnace may be offered - the primary furnace was a three storied structure with a fire chamber, a main furnace chamber, and a top story for annealing the finished products. This is in contrast to the furnace designs of northern Europe. While excavations have shown that these were not always rectangular in shape, they differ from the southern style by having their main and subsidiary furnaces on the same level. The northern designs also incorporated a different system of heat transfer by having the same fire trench running the length of a composite furnace such as Theophilus describes or else transferring the heat from the primary to subsidiary structures (Newton and Davison 1989:118-119). While originally circular in shape, the southern design was later modified after the mid-16th century to incorporate a structural addition that would allow for easier and more controlled annealing. In addition to the main furnace, there was a second, subsidiary furnace with two chambers. This *calcara* was used for the preliminary preparation of the raw materials into a frit which was later re-heated a second time and melted in the *fornasa.* Renaissance-era inventory lists of glass houses confirm the presence of two, and sometimes three, furnaces in a Muranese glasshouse.

It is interesting to speculate on the reasons that the furnace designs diverged in different parts of Europe. The southern style seen in Venice and other parts of Italy obviously is derived from those in use during the Roman era. For example, Pliny mentions small beehive-shaped furnaces. Near East influences, recorded in the 9th-century Syrian description, are also to be found. Venice has been shown to have had extensive pre-Renaissance contact with the Levantine region with respect to trade in materials associated with glassmaking like cullet and soda ash (Zecchin 1987:5-9). It would appear that the basic design of Venetian glass furnaces was a continuation of the earlier Roman form coupled with modifications seen in Levantine furnaces. The general picture of a glass furnace constructed in a compact and circular beehive shape makes sense in light of glassmaking practiced in an urban setting with limited available space, particularly that of the relatively small and crowded island community such as Renaissance Murano. Other factors that are more difficult to test such as temperature uniformity, strength, and thermal efficiency, should be considered as well.

OTHER MATERIALS ASSOCIATED WITH GLASS FURNACES

Crucibles

One of the unique features of the Renaissance Venetian glass industry is that all of the raw materials required for glass production had to be imported into the city. This includes ground quartz, soda ash, manganese, coloring agents and, as I have discussed, the materials used in furnace construction. The same observation can be made for the materials used to make glass crucibles as well as the wood needed to fire the furnaces.

One recorded source for crucible clay was the town of Valenza (sometimes called Valencia) (Jacoby 1993:78). This town was located about 60 km southwest of Milan. In addition to being used in the fabrication of clay crucibles or *padelle*, the clay was also used to line the inner walls of the glass furnace. It is known from a Milanese petition of 1394 that clay from Valenza was frequently exported to Venice for use in the glass industry. By the early 15th century, another source for a suitably refractory-type clay was Constantinople. Zecchin (1990:181) notes that the first indication of clay coming from this location dates to 1405. Presumably, only the raw clay was imported into Venice rather than finished crucibles to minimize risks of their breakage during shipping.

Inventory lists record the presence of clay from Valenza side by side with clay from Constantinople. For example, an inventory of 1446 has 4500 *libbre* of "*creda da Valenza*" (i.e. clay from Valenza) along with 12000 *libbre* of clay from Constantinople. The price of *padelle* made from the Valenza clay was slightly less than those made with clay from Constantinople (Zecchin 1990:187). This variability in price resulted most likely from either the longer distance the clay had be shipped or the better quality of the Constantinople clay. An indication of the value of glassmaking crucibles may be seen in a 1508 list in which 50 crucibles "of all sorts" were assessed at 20 ducats or about 2.5 *lire* each (Zecchin 1990:60).

It has been suggested that the crucibles used in the glassmaking process were not made by the glassmakers themselves but were rather fashioned and fired by specialist potters (Jacoby 1993:78). This supposition is based Biringuccio's text which describes how the pots are fashioned (Smith and Gnudi 1942:128-29):

The aforesaid pots or vessels are made on the wheel by master

> potters with this clay [described earlier as from Valencia] after it is well prepared. In size, they are 3/4 *braccio* high [Smith and Gnudi have suggested that this refers to the diameter], ½ *braccio* wide at the mouth and the bottom...and one *braccio* high. These are made uniformly thick, and then left to dry out gradually in the shade.

The final step of preparation appears to have been completed by the glassmakers:

> After six or eight months from the time when they were made, when you wish to put them in the furnace to begin work, that place which you left open under the arches (a hole in the furnace) is a quarter closed with a wall and only enough space is left to allow one of the said vessels to enter. Then fire is applied and continued until the whole furnace is thoroughly red-hot (about 700-900°C) and at the same time all the vessels that you wish to place in the furnace are placed in the reverberatory furnace where the frit was made and with a slow fire they are started to warm and the humidity is driven off. Then the fire is increased so that they become hot and very red.

Once the glass workers are satisfied that the crucibles can withstand the heat of the furnace without cracking, the pots were transferred from the *calcara* to the *fornasa* with special iron tools and then charged with frit and/or broken glass. Biringuccio goes on to remind the reader to ensure that the pots were not broken or leaking glass. Not only would this cause "great loss instead of profit" but it would make the pots stick to the shelf on which they rest. In addition to the very corrosive glass damaging the furnace, sticking to the shelf in the furnace would make their subsequent removal very difficult (Smith and Gnudi 1942:130). Like many of the other operations in the glasshouse, the preparing and transfer of the glass pots was an activity that required the concentrated teamwork of several workers at one time.

In comparison with the number of analyses of Renaissance Venetian glass that have been published, little compositional or structural information is available regarding the secondary materials of glass production such as the refractories or crucibles. Table 1 presents the results of analyses done on 8 Renaissance-era crucible fragments. All samples were excavated from locations in the Venetian lagoon and dated by stratigraphy and their archaeological context to the 15th and 16th centuries. Note that all analyses were done on the clay matrix of the crucible material and the results do not include the temper additions.

The bulk compositions of the crucibles vary considerably from each other possibly due to the use of different raw clay and temper types, such as crushed granite, basalt, and grog, during manufacture. This can especially seen by considering the crucible's overall amount of fluxing material (such as the percentage of $Na_2O + K_2O + CaO + Fe_2O_3$) as well as the differing amounts of SiO_2 and Al_2O_3. The crucibles have a composition that is basically comparable to that seen in modern fireclay bricks. However, the Renaissance materials typically have lower Al_2O_3 contents coupled with a much greater amount of fluxes and other oxides. This difference in composition would have served to lower the overall refractory nature of the material as compared to a modern crucible.

Table 1. Bulk analyses of Renaissance Venetian glass crucible fragments (average values in weight percents).

Oxide	1*	2*	3*	4	5	6	7	8
SiO_2	71.0	82.2	80.7	70.8	69.8	72.9	69.7	69.4
Na_2O	2.1	0.9	0.6	1.7	1.4	3.2	4.4	4.6
CaO	0.6	0.5	0.3	2.0	0.9	0.7	8.3	3.1
K_2O	3.2	2.1	1.1	3.0	1.3	2.1	2.2	3.9
MgO	0.9	0.6	0.5	1.3	0.9	0.5	3.3	1.3
Al_2O_3	20.3	12.3	14.7	19.5	14.0	16.9	9.6	15.5
Fe_2O_3	1.0	0.6	1.4	1.7	1.5	3.3	1.8	1.7
TiO_2	0.7	0.2	0.5	0.3	0.2	0.3	0.2	0.2

* Samples 1-3 were analyzed by Marco Verita of the Stazione Sperimentale del Vetro in Venice, Italy using wavelength dispersive spectrometry (WDS); see Verita, 1985. Samples 4-8 were analyzed by the author using SEM-EDS.All samples are from the 15th-16th century site of Fusina in the Venetian lagoon.

Such variability in crucible compositions has been noted by other researchers (cf. Freestone 1989). At a fundamental level, Freestone suggests that the resources available to the crucible maker were the same as those available to the potter. This must be amended somewhat in light of the fact the glass crucible makers in Renaissance Italy apparently tended to favor clays from particular locales. While it is not possible to examine samples of these raw clays, it is plausible that these materials were selected for their better-than-average refractory properties which justified their transport costs.

In addition, to compositional analyses, thin sections of four samples (# 5-8) were also prepared for the purpose of examining the microstructure and type of temper used. Microscopic observation confirmed the use of a very broad range of temper types. For example, sample #5 contained the very unusual addition of a yellow glassy material. SEM-EDS analysis and petrographic examination suggests that these particles, which are as large as 3 mm in size, are bits of slag (David Killick, personal communication, 1996); see Figure 5. While certainly added deliberately, the use of glassy temper such as this is surprising as re-firing showed that the particles began to soften at relatively low temperatures (c. 1200°C) and would have been a rather poor choice for a temper addition.

It has been noted that some medieval metalworking crucibles underwent a change in temper additions around the 14th century (Freestone and Tite 1986). The switch was from the use of quartz or mineral temper to a clay tempered with ceramic grog. The crucible fragments examined showed use of both temper types, at times in the same fragment such as #5. In all cases, the overall amount of temper added, regardless of type, was quite substantial, approaching 40% at times (based on area observation of thin sections). In addition to the "slag" particles described above, sand, basalt, granite, and grog additions were noted. Overall, the temper in the crucible fabric does not appear to have been well sorted and a wide range of sizes was typically visible.

The shape of the temper varied from well-rounded (suggesting that it was water tumbled) to other samples where the temper was very angular indicating that it was ground before being added. Figure 6 shows the crucible fabric of sample 7. It is possible to see well-rounded sand grains with adjacent cracking caused by cyclical heating and cooling. Reaction rims around the quartz particles present are also visible. Fine mullite crystals were also observed in this sample. All of the petrographic samples, except #6, had clear signs of vitrification that had occurred in the crucible fabric during its manufacture or use.

It is difficult to reconstruct the original shape of the crucibles based on the available fragments. They are all about 1.5 to 2.5 cm thick and were coated on what would have been the interior with a very thin adhering layer of glass. Glass crucible fragments excavated in the Tuscany region and dated to the 14th century may offer some parallels to the form of Renaissance-era glass crucibles (Mendera 1988:310-12). While three variations in shape are recorded, the basic template is either a bowl or truncated cone with walls 1.5 to 3 cm thick. Rim diameters are reported to be between 25 and 35 cmwhich conform to the size suggested by Biringuccio's 16th-century treatise.

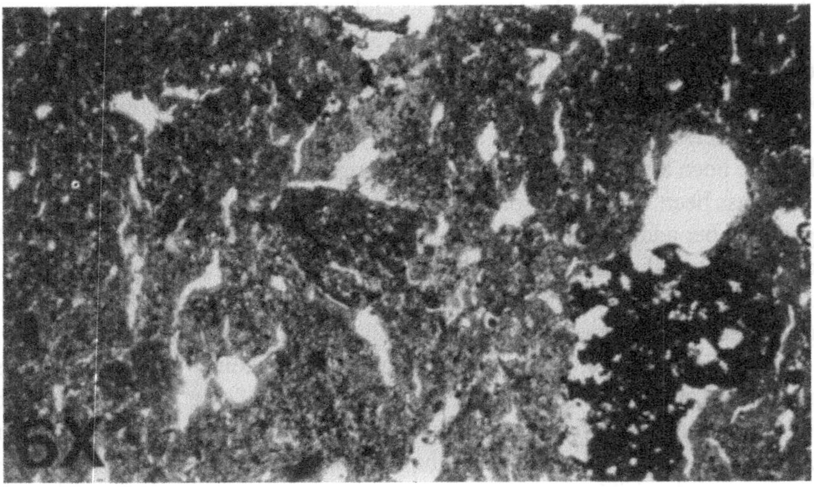

Figure 5 - Micrograph of a thin section prepared from a glass crucible sherd (#5). Note the presence of the rounded "slag" particles in the fabric.

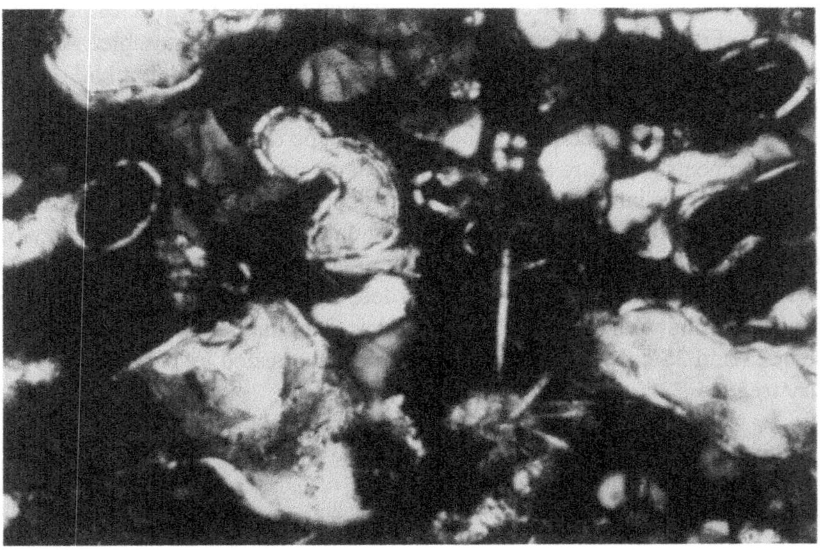

Figure 6 - Micrograph of a thin section prepared from a glass crucible sherd (#6).

Wood Supply

As with the crucibles and refractòries, little attention has been paid to oneof the most obvious and essential components of the glass industry - namely a steady supply of wood for the industry. Like all of the other raw materials, this also had to be imported into the city for use by the glassmakers. The guild rules of the glassmakers specified that only certain types of wood were to be burned in the furnaces. Originally, only alder and willow were used; this was later amended so that the furnaces burned only alder. The ability of the glass shop owners to acquire an adequate supply of wood does not, from the archival information at least, appear to have been a major problem. This is despite the fact that the renewed production of luxury glass in Venice after 1450 would have increased the consumption of wood greatly. This is not only due to the greater number of shops in operation but also the lengthier processing and preparation steps required for making *cristallo* and other luxury glasses.

There are only a few sources for wood described in the archival material. A notation from 1331 says that wood was obtained from the area near Cervignano and transported by water to Venice (Zecchin 1987:16). This town was about 80 km northeast from Venice. Other references from the mid-15th century note the towns about 50 kilometers northeast of Venice as additional sources (Zecchin 1990:180). Some disputes among glassmakers regarding wood supplies are seen. These do not appear to indicate a chronic shortage of materials but rather specific disagreements about distribution.

The various shipping lists from the 14th and 15th century published by Zecchin give an indication of the overall amount of wood consumed in Venice's glass factories in a given year. This consumption has been shown to have increased notably after ca. 1450 (McCray 1996). For example, in 1447 wood consumption was some 4070 *carri* (1 carro was equal to about 1 cubic meter of wood) according to the receiving lists in the archives (Zecchin 1990:51). By 1455, this had increased to about 7800 *carri*. The amount of wood received by any one shop in Venice varied depending on their degree of output and the type of glass objects they were fashioning. The amounts requested by individual shops in 1455 ranged anywhere from Taddeo Barovier's shop (1340 *carri*) to Antonio Parisi's furnace (45 *carri*). The 1508 inventory of the Dragani family workshop does show that wood was a substantial part (about 22%) of the overall value of the shop (Zecchin 1990:60).

INFERENCES REGARDING OPERATING PARAMETERS

Michael Polanyi (1962:328) uses a concept he calls the *operational principle* of a device (such as a glass furnace). This means the manner in which the parts of the device fulfill their special function in an overall operation which achieves the purpose of the device - in short, how it works. This definition also incorporates the purpose of the device. What was the primary function of the Renaissance glass furnace? Depending on the type of furnace considered, the functions of the furnace varied from heating the raw materials of glassmaking at an intermediate temperature in the preparation of a frit to the prolonged and extensive heating of this frit to promote melting and homogenization of the glass. Other parts of the furnace had the function of annealing the glass after it was manufactured.

On the basis of documentary information and materials analyses it is possible to make some inferences regarding the manner in which glass furnaces, especially the primary ones where the glass frit and cullet were melted, were operated. The three primary variables that are involved with the preparation and melting of glass prior to its working are the atmosphere of the furnace, the time duration of fritting, melting, or homogenization, and the temperatures employed. Proper supervision and monitoring of these variables was extremely important especially in the successful production of Venetian luxury glass such as *cristallo*.

The first furnace operation was the preparation of the frit. This operation was carried out in the first furnace (the *calcara*). An intermediate heat (about 700-800°C) was applied to a mixture of the crushed quartz and soda ash. This was done for several hours and the mixture was carefully monitored and continuously stirred to avoid the formation of a liquid (Verita 1991:58). Neri states in his text (Chapter 2, Book 1):

> ...put them [the raw materials] into the calcar which at first must be well heated...at first for an hour make a temperate fire and always mix the frit with the rake that it may be well incorporated and calcined; then the fire must be increased always mixing the frit with the rake for this is a thing of great importance and you must always do this for five hours, still continuing a strong fire.

The *calcara* had to be heated prior to introducing the raw materials due to the insulating character of the finely powdered mixture (Turner 1963:189). The end product of the fritting process was a solid mass which was broken up and stored in a clean, dry place. The importance of a "temperate fire" appears in more of Neri's recipes as well. The formation of a liquid would be bad in several senses. As the calcination of the raw materials was done on the shelf of the *calcara* and not in crucibles, the formation of a liquid would result in the loss of material. A liquid material would be much harder to handle and it would have a very corrosive effect on the furnace itself all while absorbing impurities from the refractory material (Turner 1956b).

While the events taking place inside the fritting furnace would depend on a combination of factors such as the furnace atmosphere, purity of the raw materials and so on, a general picture can be presented. The main constituents present would be SiO_2 along with Na and K in the form of carbonates, sulphates, and chlorides. Reactions among these materials are slow until 800- 900°C. The solid material remains as a powder below 700°C and only begins to assume a sintered or fritted state around 750° C. The melting point of Na_2CO_3 is 851°C and that of NaCl is 801°C. Therefore to avoid forming a liquid the temperature would have to be maintained around 750°C and kept there for some time to allow time for the reactions among the various constituents to occur (Turner 1956b:292). At this temperature, little loss of NaCl could be expected and any potassium and sodium present in the form of sulphates would probably also be unaffected. As a result, most of the chlorides and sulphates would pass over to the glass-melting crucible unchanged.

Once the frit had been prepared, it could then be melted. This process was carried out in the main furnace. The frit was mixed with cullet, usually, to aid in the melting process (Verita 1989:60). It would then be heated to a much higher temperature, probably around 1100-1200°C. Based on the study of model glasses (75%SiO_2-10%CaO-15%Na_2O), the additional reactions taking place in the melt can be described. In the melting process, the alkali salts would liquify, dissolve into one another, and take calcium and magnesium carbonates into solution. These would rapidly decompose and reactions between K_2CO_3 and Na_2CO_3 with SiO_2 would occur (Turner 1956b:294-95). The chlorides and sulphates were still present after the fritting step. These have limited solubility in the glass. For example, the solubility of NaCl in the model glass as reported by Turner was 2.3% at 1400°C. The solubility would be further decreased at the lower temperatures attainable in the Venetian furnaces. Some of these impurities were removed by skimming the surface of the glass melt with a special tool.

The next basic stage in production was the refining and homogenizing stage. This was carried out by maintaining the furnace at the highest temperature possible for a long period of time. Neri notes that the should be melt "should be let to boil 4 to 6 days" (Chapter 9, Book 1). The recipe for glassmaking in Biringuccio's work prescribes a time of about two days (Smith and Gnudi 1942:130). While the glass is being heated and refined, several different steps are carried out. For example, the melt as a whole was continually stirred.

The purpose of this stirring was to eliminate the different defects that can occur in the glass, as a material, during the melting process. Two of these possible defects are bubbles (very small bubbles are called seed) and cord (striae is the term given to finer sized cord). Bubbles become trapped in the melt due the evolution of gases during the melting process. This would have especially been a problem as the Venetians used a sodium-based fluxing agent made primarily of carbonates. During the heating and melting process, CO_2 gas would have evolved. While air bubbles, in time, can rise to the surface of the melt and escape, this speed of this process depends on their size and the viscosity of the glass, among other factors. Stirring the melt continually would have helped eliminate these by bringing them to the surface where they could escape. Cord and striae result from a poor mixing of the glass melt. These can be seen in the finished product as wavy striations or streamers. They are essentially caused by compositional differences in the glass resulting in localized regions having a different index of refraction. Stirring the glass melt over a period of time would have helped homogenize the composition of the glass and allow it to become more uniform. Eliminating these inhomogeneities gave a glass with greater clarity and a more uniform appearance. Because the glass furnaces used in Venetian production were not capable of attaining the same high temperatures of modern, or probably even 18th-century coal-fired furnaces, the homogenizing and refining of the *cristallo* glass melt through continual and sustained stirring was an integral part of successful production.

As the glass was stirred, it was continually monitored by a specialized employee called *conciatore*. This was the person responsible for activities such as preparing the glass batch in the proper ratios. Another specialized job in the Renaissance Venetian glass furnace was the *stizador* who was responsible for stoking and running the furnace (Zecchin 1987:7).

Neri notes towards the end of Chapter 9 of Book 1 that additional manganese may be added on the judgement of the *conciatore* if the glass still had a greenish color due to iron impurities. Also, during the entire melting and refining process, it was important to maintain the proper furnace conditions. An anonymous 16th-century recipe book from a private Venetian collection instructs that the furnace was fired so as to have bright white flames and to be free of smoke or soot which could cloud the glass (Moretti and Toninato 1987:33). Neri gives similar directions in his original 1612 version which were not copied directly in the English translation. Again, the furnace is kept hot and stoked with "strong and dry wood" (Barovier, 1980:13). One is also to guard against fumes; Merrett's version of Neri's work has comments at the end offered by the translator and essentially confirms the importance of maintaining these furnace conditions. In addition to keeping the furnace as hot as possible, all of the instructions note that the furnace should be free of smoke and fired with dry wood. If carried out, these instructions would have created a fairly oxidizing atmosphere inside the furnace. This would have helped drive the reaction between the iron and manganese:

$$Fe^{+2} \quad + \quad Mn^{+3} \quad \rightleftarrows \quad Fe^{+3} \quad + \quad Mn^{+2}$$

to the right. In this manner, even more of the greenish-blue tint caused by the Fe^{+2} ions could be eliminated giving a more color-free glass.

From the preceding discussion, it can be seen that the successful Renaissance production of a fine glass melt, such as *cristallo*, suitable for working into luxury glass, was a complicated and lengthy process involving a great amount of judgment and tacit knowledge. It incorporated the systemic use of steps beginning with the careful selection of the proper raw materials and concluding with an attentive eye on the furnace conditions. Obviously, there were many places in this chain where problems could arise. Therefore, the ability of the Muranese glassmakers to produce quality and relatively colorless glass must described as a "team effort" requiring informed decisions on the part of everyone from the *stizador* who watched the furnace to the *conciatore* who prepared the glass batch.

What temperature were the primary (i.e. melting) glass furnaces used in Renaissance Venice capable of attaining? One answer to this question can be found by again considering their function -- they had to be capable of heating a glass melt to a temperature such that it could be gathered on a blowpipe and blown before being worked. Gathering and blowing are glassmaking operations that are typically carried out when the glass has a log viscosity of about 3 to 4 (Brill, 1988:280). Based on compositional analyses of the glass made in Renaissance Venice and the use of existing models for glass viscosity found in the materials science literature (Lakatos, 1972), it was possible to make a plot of estimated temperature versus log viscosity for a typical Venetian glass. An example, in this case for a typical Venetian *cristallo* or *vitrum blanchum* glass, is shown in Figure 7. From this, one can see that a log viscosity of 3 to 4 is attained in the temperature region of about 1100° to 1200° C. This implies that the glass furnaces of Venice would have to have been able of achieving temperatures in this region in order to perform their technical function. From the same plot, one can see that annealing operation (typically done when the glass has a log viscosity of about 11 to 13) was probably carried out a temperature of around 450°-600°C.

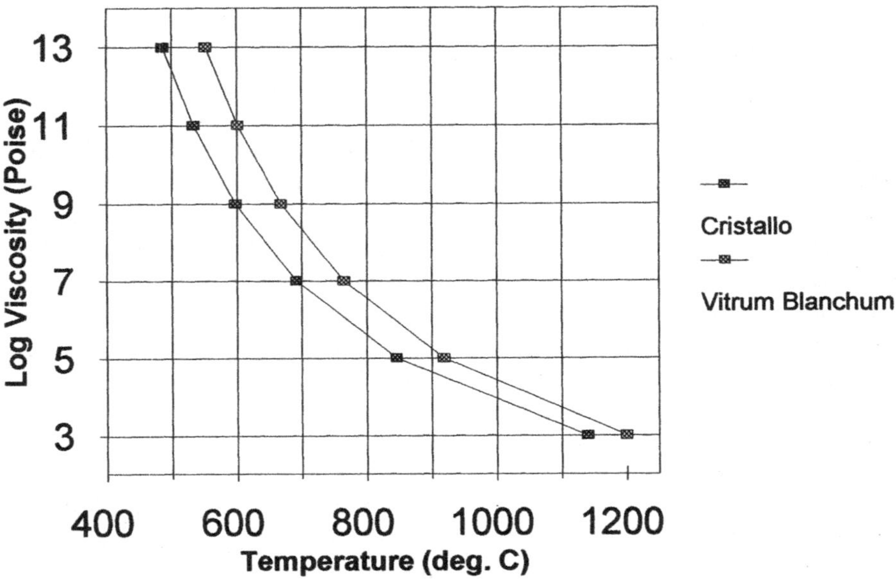

Figure 7 - Hypothetical viscosity-temperature plot based on average Venetian *cristallo* and *vitrum blanchum* glass compositions.

The inferences regarding the maximum possible temperature that could be achieved in the Renaissance-era glass furnace based on viscosity-temperature models were essentially confirmed through the use of simple re-firing tests. Small fragments of Samples 5 and 6 were heated at atmospheric conditions at 50°C temperature intervals from 1100° to 1350°C in an electric furnace. The visual differences noted between the samples heated to different temperatures are shown in Table 2. The results suggest that the maximum temperature experienced by these particular fragments was most likely between 1150° and 1250° C.

Table 2. Results of re-firing tests done on samples 5 and 6.

Temperature (°C)	Sample 5	Sample 6
1100	No visible change	No visible change
1150	No visible change	No visible change
1200	Partial melting of temper additions	No visible change
1250	Melting of temper and noticable pore formation	Noticeable color change; some slight surface"glazing"
1300	Slumping and melting	Noticeable surface glazing
1350	Complete melting	Melting, bloating, and slumping of fragment

CONCLUSIONS

The basic design of the southern versus the northern furnace appears to be well-understood. Ideally, however, more attention should be paid to the analysis of refractory materials as they are excavated from glassmaking sites. While I have presented inferences regarding operating variables, it would be interesting to see whether these bear the scrutiny of additional studies. Also, additional attention to the question of glassmaking crucibles would be welcome. For instance, can any general pattern be discerned in either the selection of temper types used in glassmaking crucibles and other refractory materials as they vary in time and location as compared to other ceramic materials? What does this say about the nature of specialized raw material selection?

To conclude, a number of different sources of information, including archaeological and documentary material, pictorial representations, and material analyses, have been presented to give a holistic view of Renaissance glass furnace technology. The primary focus of the presentation was on glass furnaces as constructed and used in Murano as part of the Venetian tradition of glassmaking.

The southern style of glass furnace can be seen to have gradually changed over time from the simple "beehive" structure referred to by Pliny to one that was enlarged and modified to include a tunnel *lehr* for more controlled cooling. At the same time, the basic template of a round furnace shape was retained through time.

The construction and operation of Renaissance glass furnaces, especially those of Venice, involved specialization and selectivity with respect to the materials used and the division of labor associated with running it. While not discussed in detail here, there was also considerable specialization with respect to the tools used in fritting, homogenizing, and melting the glass. Furthermore, all of the raw materials related to glassmaking in Renaissance Venice had to be imported into the city including refractory materials and wood.

In addition to the main melting furnace, documentary and pictorial sources indicate that there were other smaller furnaces employed in glassmaking operations, each with their own specific functions. The main furnace was where the melting, homogenizing, and working operations were carried out. Therefore, this structure had to serve several functions including providing a space for the finished glass to be annealed. One of the advantages of the southern style of glass furnace may be seen in that it arguably would have occupied less space in the crowded urban environments where they tended to be located.

Finally, some attention has been paid to the operating conditions of Renaissance glass furnaces such as temperature and atmosphere. Glassmaking was a team-oriented activity and careful attention to these variables on the part of the furnace staff (such as the *stizador* and the *conciatore*) was an important part in the successful manufacture of the technically demanding luxury glass compositions which were the hallmark of Renaissance Venetian glass production.

ACKNOWLEDGMENTS

The author wishes to thank the organizers of this symposium for an enjoyable and informative meeting. The program on Culture, Science, Technology, and Society at the University of Arizona provided travel funds for which the author is grateful. David Johnson and David Killick at the University of Arizona assisted with the electron microscopy work and thin section examination. Ernesto Canal provided the crucible fragments studied.

REFERENCES

Barovier, R. 1980. Antonio Neri's L'Arte Vetraria, Milan:Polifilo.

Brill, R. 1988. "Scientific Investigations of the Jalame Glass and Related Finds" in G.D. Weinberg, ed., Excavations at Jalame: Site of a Glass Factory in Late Roman Palestine. pp. 257-295. Columbia: University of Missouri Press.
Calegari, M. and Moreno, D. 1975. "Glass Production in Liguria from the Fourteenth to the Sixteenth Century", Archeologia Medievale, 2:13-29.
Charleston, R. 1978. "Glass furnaces through the ages", Journal of Glass Studies, #20:9-44.
Freestone, I. 1989. "Refractory materials and their procurement" in Old World Archaeometallurgy: A Conference Proceedings from 1987, Bochum, 155-162.
Freestone, I. and Tite, M., 1986. "Refractories in the Ancient and Pre-Industrial World", in High Technology Ceramics: Past, Present, and Future W. D. Kingery (ed.),pp. 35-63. Columbus: American Ceramic Society.
Heikamp, D. 1986.Studien zur Mediceischen Glaskunst, Florence: Kunsthistorisches Institut.
Gasparetto, A. 1967. "A Proposito dell'officina vetraria Torcellana", Journal of Glass Studies, #9:50-75.
Jacoby, D. 1993. "Raw materials for the glass industries of Venice and the Terraferma about 1370-about 1460", Journal of Glass Studies ,#35:65-90.
Lakatos, T. et. al.. 1972. "The effect of some glass components on the viscosity of glass", Glastek. Tidskr. 27 (2):25-28.
Månsson, P. 1520. Glaskonst (translation by R. Geete), Stockholm.
McCray, W.P. 1996. unpublished dissertation research.
Mendera, M. 1988. "Some Aspects of Medieval Glass Production in Central Italy" in Annales du 11th Congres de l'Association Internationale pour l'Histoire du Verre, Liege, 303-315.
Mendera, M. (ed.) 1991. Archeologia e Storia della Produzione del Vetro Preindustriale, Florence: All'Insegna del Giglio.
Moretti, C. and Toninato, T. 1987. "Cristallo e vetro di piombo da ricettari del '500, '600,' 700", Rivista della Stazione Sperimentale del Vetro(1):31-40.
Neri, A. 1662. The Art of Glass (as translated by C. Merrit), London.
Newton, R. and Davison, S. 1989. The Conservation of Glass, London: Butterworths.
Polak, A. 1975. Glass: Its Tradition and Makers, New York: Putnam.
Polanyi, M. 1962. Personal Knowledge, Chicago: University of Chicago Press.
Redi, F. 1991. "Una Vetraia Rinascimentale" in Archeologia e Storia della Produzione del Vetro Preindustriale M. Mendera, ed., pp. 91-98. Florence: All'Insegna del Giglio.
Smith, C. and Gnudi, M. 1942. The Pirotechnica of Vannocio Biringuccio, New

York: Inst. of Mining and Metallurgical Engineers.
Tabacyznska, E. 1968. "Remarks on the Origins of the Venetian Glassmaking Centre" in Studies in Glass History and Design, R. Charleston, ed., pp. 20-23. England: Greshham Press.
Tabaczynska, L. et. al.. 1977. Torcello Scavi 1961-62, Rome.
Tait, H (ed.). 1991. Glass: 5000 Years, New York: Abrams.
Turner, W. 1956b. "Studies in ancient glasses and glassmaking processes part V: Raw materials and melting processes", Transactions of the Society of Glass Technology, 40:277-299.
Turner, W. 1963. "The Tercentenary of Neri-Merrett's *The Art of Glass*"in Advances in Glass Technology 2, 181-20.
Verita, M. 1985. "L'invenzione del cristallo muranese: una verifica analitica delle fonti storiche", Rivista della Stazione Sperimentale del Vetro, (1):17-29.
Verita, M. 1989. "Some Technical Aspects of Ancient Venetian Glass" in Technique et Science les Arts du Verre, Namur, 57-67.
Zecchin, L. 1987. Vetro e Vetrai di Murano I, Venice: Arsenale Editrice.
Zecchin, L. 1989. Vetro e Vetrai di Murano II, Venice: Arsenale Editrice.
Zecchin, L. 1990. Vetro e Vetrai di Murano III, Venice: Arsenale Editrice.

UNUSUAL OLD KILNS

O. J. Whittemore, 10015 Lakeshore Blvd. NE, Seattle, Washington 98125

ABSTRACT

Several old kilns, some of unusual construction, are described including the Chinese hill-climbing kilns, a Hoffman ring kiln, a continuous chambered kiln, 2 vertical (!) tunnel kilns, a pebble-bed regenerative kiln for 1750C and a rotating tube kiln for thermocouple protection tubes.

INTRODUCTION

The firing of ceramics is probably the most important step in processing. Many kilns have been designed and constructed for firing. The evolution of kilns has been determined by labor cost, energy cost, product quality, and capital cost after the temperature, atmosphere, and volume required have been determined.This paper will describe several unusual kilns (some of which were fired by the author), some which were used significantly, some abandoned.

The earliet kiln was the open-top scove kiln. Dried bricks were fired by stacking them to provide a solid outer casing, spaces underneath for the fire, and spacing between the upper bricks for passage of the combustion gases. After firing, the outer poorly fired bricks were set again in later firings. These were the kilns used by ancient clayworkers who were commanded in the Bible, "Go to, let us make brick, and burn them thoroughly". (Genesis 11:3). Scove kilns were used into this century and billions of brick have been fired in them. The low product quality and the high labor costs of setting and drawing have made these kilns obsolete.

Some of the earliest significant kilns with permanent walls were the Chinese hill-climbing kilns. A tunnel was built up a hillside and layered with refractory mud . Greene reported that these kilns lasted only one or two firings and never over 10.(1) He wrote that the debris of old kiln parts, broken saggars and rejected ware forced reconstruction at new sites. A similar kiln in Korea was described by Gilbertson(2). Pictures of these kilns are in both Bulletin articles.

A later version was cconstructed with a series of joined chambers instead of one tunnel yet still climbing up a hill. One such kiln is shown in Fig. 1, a drawing by Prof. W.J.Sutton of Alfred University(3). These were wood fired, used for porcelain, were about 4 m. high inside the chambers. After loading all chambers, the fire was started in the small chamber at the bottom. Combustion gases passed through the bottom and then in a down-draft fashion through the chambers much like the later continuous chambered kilns of Europe and the United States. (One wonders if the latter were not patterned after the Chinese kilns.) In one firing, the temperature reached cone 14 down (approximately 1400 C.) The Tehwa area was making white porcelain during the Ming dynasty and therefore for at least 500 years and probably earlier.. Pottery families owned individual chambers of the kiln and the kilns were only fired every few months. Norton also describes a hill-climbing kiln with an illustration(4).

Fig. 1 Tehwa Chinese Pottery Kiln (cross-section).

Ellis Lovejoy, a Past President of the American Ceramic Society, wrote the book, "Burning Clay Wares" in 1920(5), in which many old kilns are described from open-top kilns to up-draft, down-draft, up- and down-draft, continuous (ring, chambered, and car tunnel kilns), and muffled kilns. Most of these do not exist today. Most were fired with coal, an unusual kiln fuel today, having been replaced by natural gas, oil, or electricity.

A common kiln in the structural clay and refractories industries was the down-draft kiln (sometimes called the bee-hive kiln). Many varieties were designed and at least 12 versions differing mainly in their floor designs are described in Lovejoy's book. One such kiln is shown in Fig. 2. From the side fire boxes (10 shown in Fig. 2), the flame rose over the product , down through the product and out the floor.

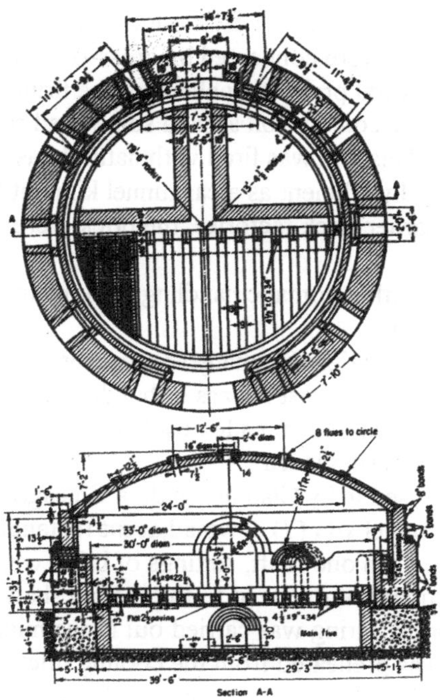

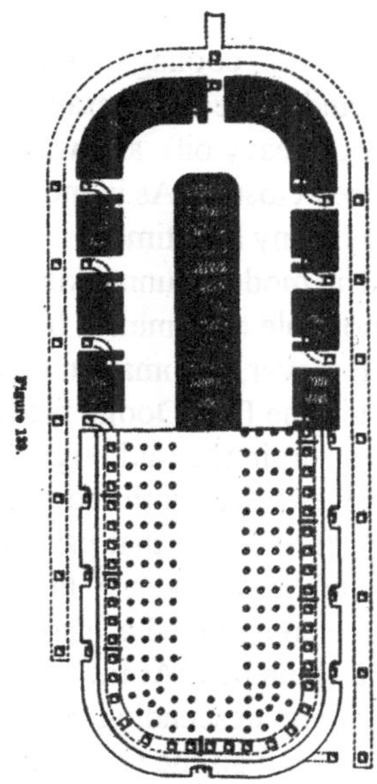

Fig. 2. Round down-draft kiln. Fig. 3. Hoffman ring kiln (plan view).

The Hoffman ring kiln (patented in 1858) was widely used for structural clay products at one time because of lower fuel cost. This kiln was an oblong annular tunnel as shown in Fig. 3 with several entrances to permit setting in one zone, firing in another, and drawing in another. Separation of zones in one case was by constructing a paper wall which burned when the firing reached that point. One such kiln in Seattle, Washington was fired with sawdust poured down into the kiln. A much smaller version is now in operation in Brazil for firing magnesia refractories to about 1700 C.

At one time a number of continuous chambered kilns were in operation which have some similarities to the Chinese chambered hill kilns. These were a series of rectangular down-draft chambers linked together and usually connected to two central flues, one for exiting combustion gases, the other for drawing hot air from cooling chambers for drying. About 12 designs are described by Lovejoy. In one such kiln designed by the inventor, Goldner, at Fort Dodge, Iowa , 9 chambers were on each side of the flues and were connected by cross-over flues at the ends. One chamber would be firing, several preheating ahead of the firing chamber, several cooling and preheating the combustion air and some used for dryer air, and some being drawn and set. This kiln was fired with natural gas (earlier with heavy oil) and was as thermally efficient as a car tunnel kiln but labor costs caused closure. As in most of the earlier brick factories, the brick were handled as many as 5 times.

The modern tunnel kiln together with automatic handling equipment has made it possible to eliminate human handling .

However, automation means loss of variety for many shapes were produced in the Fort Dodge factory including at least 3 sizes of building blocks, cove bases, drain tile (although this product has been made obsolete by plastic pipe), besides brick, both red and salt-zinc flash glazed.

Heat rises and that principle was used in by Claude Smith, a Chicago inventor and brickmaker, in his vertical tunnel kiln (7). Two adjacent rectangular shafts were constructed as shown in Fig. 4. In one shaft, settings of brick were lifted by a chain belt. At the top, the settings were moved to the second shaft on top of a slowly descending column of brick. Firing was carried out in the middle of the second shaft so the combustion air was preheated from the cooling bricks below and combustion gases preheated the column above. Finished bricks were removed at the bottom. One of these kilns was built at Sheffield, Iowa but the company did not survive a depression of the 1920s. Although the patent claims drying as a function of the kiln, I remember drying tunnels adjacent to the kiln. Kilns are not good dryers due to the water vapor from combustion in their atmospheres. (During the discussion following the presentation of this paper, a comment stated that vertical kilns similar to the Smith kiln are now used in China but only with one vertical tunnel).

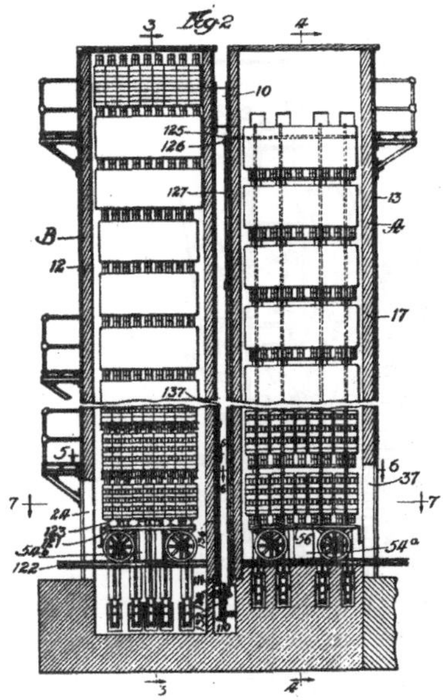

Fig. 4. Smith vertical tunnel kiln (for bricks).

In Italy, a vertical tunnel kiln was built in Torino for firing dinnerware(8). The kiln was 9.3m. high and had a 0.3m. circular inside diameter and was heated electrically to 1050 C. A slowly descending column of saggars containing the ware was lowered during operation. Much floor area was conserved by this design.

More recently,a regenerative kiln of about 2 m. inside height was designed by S. S. Kistler, research director at Norton Co., for firing pure oxide refractories to 1750 C. (Fig. 5). The bottom of the kiln was a transfer car which could be withdrawn for drawing and setting. The bottom of the car contained two zones of alumina pebble beds. In operation, air was blown through half of the pebble beds and thus was preheated. The air then mixed with natural gas flames inserted via water cooled burners. The combustion gases moved through the kiln and out through the other half of the pebble beds in the car bottom, thus transferring the heat. The air and the natural gas were reversed every 4 minutes from one side to the other. The waste gases were withdrawn through sheet metal ducts at 200 C. or less so the gradient in the 40 cm. bed was about 40 C./cm.

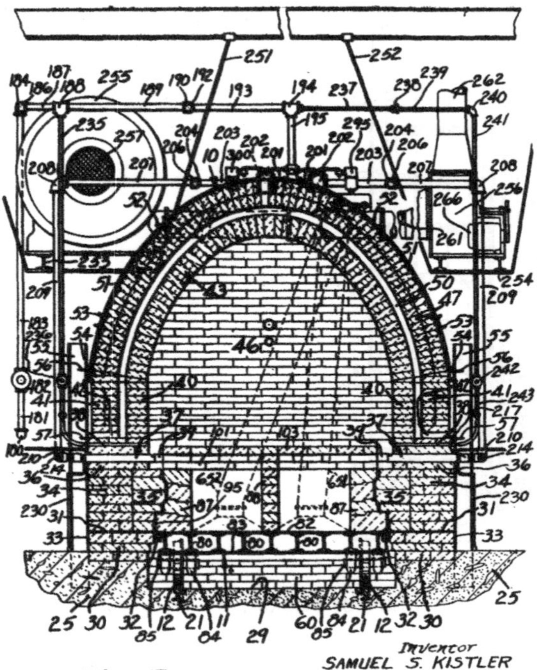

Fig. 3

Fig. 5. Kistler 1700 C. regenerative kiln.

About 40 years ago, an interesting kiln was built for firing thermocouple protection tubes and a sketch from memory is shown in Fig. 6. The kiln was about 1 meter long and was heated electrically with molybdenum elements to about 1700 C. Hydrogen was inserted into the kiln shell to protect the elements. At each end of the 8 cm. inside kiln diameter, a short rotating tube was inserted part way into the kiln. These were rotated at 6 rph (note the "h"- the kiln was first constructed for 6 rpm which broke up most of the rotating parts!). Through the rotating tubes and through the kiln were pushed porous fused alumina carrier tubes, about 6 cm. outside diameter by 1 meter long. These carrier tubes were closed at one end and tapered to fit the preceding carrier tube. They were pushed through the kiln at about 1.3 m. per hour.

Thermocouple protection tubes which had been low fired were placed in the carrier tubes and thus were rolled slowly while being fired. Tubes of one diameter could be inserted in larger diameter tubes for firing and this nesting could be used with 3 sizes. Although starting out rather warped, the tubes came out very straight. In fact, daylight could be seen through a 1.5 mm. hole in a one meter tube!

I mention this method of firing as possibly of much value although unfortunately it was not applied commercially due to a management decision. For example, perhaps urania nuclear fuel rods could be fired in this manner. To fire straight tubes previously, it was necessary to hang them which also causes problems.

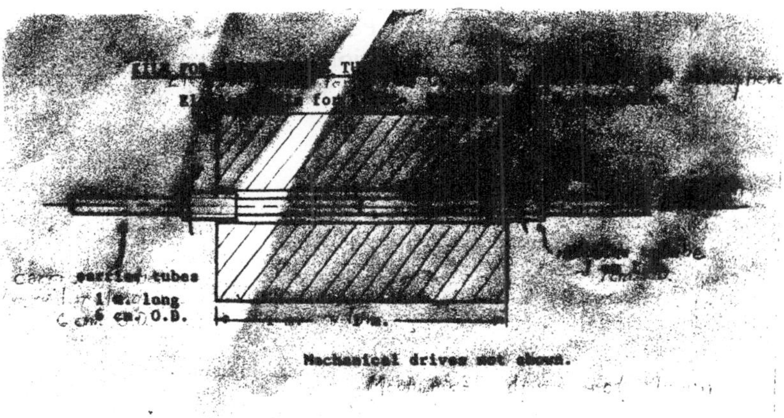

Fig. 6. Rotating tube kiln for 1700 C.

These are but a few of the kilns that have been designed in the past some proving of value, others abandoned because of obsolescence due to poor product quality , high associated labor costs, etc. Some of these designs may suggest new designs for future kilns.

REFERENCES

1. H. Greene, "Chinese Kilns That Last for Two Firings", Am. Cer. Soc. Bull. 15(9)223(1936).
2. W. Gilbertson, "Ancient Pottery Techniques in Modern Korea", ibid 29(8)297(1950).
3. W. Sutton, "The Manufacture of Porcelain Near Tehwa, South China", ibid 17(11)450(1938).
4. F. Norton, "Refractories, 3rd Ed.", McGraw-Hill, New York (1949).
5. E. Lovejoy, "Burning Clay Wares", T.A. Randall & Co., Indianapolis (1920).
6. O. Whittemore, "Iowa Structural Clay Products Industry", Cer. Age, Oct.(1941).
7. C. Smith, "Continuously Operated Kiln", U.S. Patent No. 1,578,559 (1926).
8. E. Humbert, Torino, Italy. Personal communication. (1960).
9. S. Kistler, "Regenerative Kiln", U.S. Patent No. 2,571,897.

CONCLUDING COMMENTS

Prudence M. Rice
Department of Anthropology
Southern Illinois University
Carbondale, IL 62901

The papers originally presented in the symposium on "The Prehistory and History of Ceramic Kilns" at The American Ceramic Society and subsequently gathered here stimulated all of us, organizers and participants alite, to rethink many issues of prehistoric pottery production organization and technology. For example, in 1987 I summarized what was then (I believed) a reasonably accurate view among archaeologists concerning early kilns and firings:

> Pottery is fired by two fundamental techniques: with and without a kiln. Although kilns were widely used throughout the Old World in antiquity, in the New World they had a very restricted distribution, primarily in highland and Gulf coastal Mexico ... until the arrival of Europeans and European technology in the early sixteenth century. (Rice 1987:153)

This scenario needs to be greatly revised, for, as is abundantly clear from the papers in this volume, prehistoric kilns have been found to have been used fairly widely in the New World, not solely in Mexico. This circumstance is forcing new attention to prehistoric developments in pyrotechnology and to geographical and temporal variations in construction and firing techniques.

EVOLUTION OF KILNS

We can begin with consideration of one very influential potter's perspective on the prehistory and history of kilns:

> Like many other devices the kiln was not an invention but rather the result of a series of small improvements. The fact that pit firing works quite well probably means that for the first several milennia of pottery making there was little motivation for change ... potters felt little compulsion to introduce even the most obvious improvements. (Rhodes 1981:13).

Rhodes (1981:3) believed that "... early pottery was fired in open bonfires or in shallow pits holding the fire, and not in kilns .... kilns, even rudimentary ones, are a very recent development ...." His scenario indicates that the most common pre-modern, non-industrial firing strategies were either "open" or "non-kiln" (bonfire, pit, clamp, i.e., lacking a formal, permanent structure) or by means of a simple structure, typically an updraft kiln, usually single-chambered and open-topped. This developmental sequence runs as follows (Rhodes 1981: 3-27):

1: first, use of an open bonfire
2: then, shallow pits - advantages: heat retained by insulating barrier of earth, thus a higher temperature; better control of temperature and atmosphere; slower cooling (less breakage); disadvantage - sudden thermal stress of rapid heat rise; uneven heating; high percentage of firing loss
3: next, use of low mud walls at sides of pit - advantages: a step toward a fixed structure; confined the fire; slowed cooling
4: placement of fire underneath ware setting (and addition of firemouths) - advantages: introduced circulation of heat (rather than just radiation); temperatures more even; fuel more efficiently used; closer step toward an enclosed structure
5: covering the top of the ware - contains heat, decreases radiant heat loss upward; disadvantage - uneven heating
6: creation of a separate chamber for fuel below ware chamber - advantages: pots not subjected to direct flame; more controlled heating and circulation; fixed structure capable of multiple use
7: final step, placing a roof over the kiln.

This evolutionary trajectory is admittedly appealing in its simplicity, but at the same time it is painfully (to anthropologists) reminiscent of Lewis Henry Morgan's (1964) late 18th-century unilineal evolutionism, in which he identified a series of material markers for human progress from "savagery" through "barbarism" to "civilization." The lack of unequivocal evidence for this sequencing--even its very premise--means that archaeologists, potters, and ceramic engineers must work more closely together to try to answer questions about why particular firing strategies developed, and when. Questions about the reasons for innovations and inventions in prehistoric technologies have puzzled archaeologists for decades, and although no definitive answers have been found it is clear that we are still a long way away from really understanding "motivations" and "compulsions" for prehistoric change. Our late 20th-century bias toward technical progress and improvements in "efficiency" (i.e., cost-effectiveness of expenditures of time or fuel) may act as a kind of filter that actually obscures our vision, rather than clarifies it.

For example, my earlier comment on kiln distributions, which I quoted above, was followed by a statement that "Kilns probably were an early response to three needs of potters: more control over products, higher firing temperatures, and more economical use of fuel" (Rice 1987:153). Permanent kiln structures are widely believed to offer striking advantages for potters in firing their wares as compared to open firings. Kilns are said to allow higher temperatures (ca. 100° C higher), separation of fuel and ware, more effective heat containment, greater volume of production, more possibilities for different ware settings of ware, and better control of temperature, rates of heating, and temperature uniformity. The disadvantage of permanent kiln structures is that some of the heat generated by combustion has to go toward heating the structure itself, and there may be a longer time of firing (i.e., a slower increase in temperature). But is this necessarily a disadvantage? Are these views borne out by the contributions to this volume?

Some authors talked to traditional potters near their sites about their use of kilns: What do potters claim is the advantage? One potter that Christopher Pool interviewed stated that the advantage was better control of the rate of heating, specifically a slower rate of heating. Pool suspects that control of the rate of heating was particularly important for manufacture of fine-paste wares, and that control of firing atmosphere was also significant. His interpretation closely follows the observations of Gossselain (1992), who compared five different firing strategies (open, pit, open with cover sherds, pit with cover sherds, updraft kiln) on a number of criteria. Gosselain (1992:257, emphasis added) concluded that "The only parameters allowing a technological differentiation [among firing techniques] are *the heating rate* and ... the time of exposure to temperatures" (i.e., soak time). Significantly, firing strategies could not be differentiated on the basis of temperatures, because all reached overlapping temperature ranges of 600-900° C.

Was fuel conservation a concern? While several authors addressed questions of fuel quality and quantity, and possibilities of the exhaustion of wood fuel sources in prehistory, quantity and conservation seems to play a negligible role. Fuel use for pottery firing was probably minimal compared to household use for cooking, and in some areas other crafts such as metalworking would have been comparable consumers. Quality of fuel seems equally important as quantity.

WHAT IS A KILN?

We can return to some of the issues brought forward in the Introduction to this volume, and approach kilns from two pragmatic perspectives, physical and functional. Daniel Rhodes (1981:21) has discussed three physical components of kilns: (1) the fireplace (firebox) or mouth (firemouth) to burn fuel and generate heat; (2) the chamber(s) to hold ware and retain heat; and (3) flue(s) or an exit for escape of hot gases and creation of a draft to move heat through the facility. David Kingery (this volume) presents us with a complementary set of four functional principles that define a kiln by what it does: (1) generate heat, (2) contain heat, (3) transfer heat, and (4) hold the ware. It is clear from Kingery that "enclosure" is a key element in two functions of kilns, containing and transferring heat. What must be highlighted here are the relations between and among Rhodes' components and Kingery's operating principles, for these will determine the physical layout and functioning of a kiln and aid archaeologists in interpreting what are sometimes ambiguous remains.

And again we can ask, What is a kiln?, and try to modify the answer to make it more relevant in archaeological and prehistoric contexts. I earlier offered the definition of kilns as "enclosed chambers for containing and channeling combustion in the firing of ceramic materials" (Rice 1987:158). Here it is useful to remember that the industry definition given in Chapter 1 adds the notion of permanence by referring to a kiln as an "installation." As seen in some of the papers (and elsewhere), however, simple round updraft kilns of adobe are seldom "permanent," because firing stresses and other circumstances often require that they be reconstructed after 3-20 or so years of use. Firing without a relatively permanent "installation" or an enclosure is usually called "open firing" or "bonfire" firing. I earlier considered "pit kilns" to be "functionally intermediate between open firings and updraft kilns in that they involve enclosures" or (Rice 1987:158; this type of firing has also been called "mixed firing". Rye and Evans 1976:164-166]). Clearly the "trench kilns" of the Anasazi area (Blinman and Swink, this volume) are a subset of pit kilns. But now a new wrinkle has appeared with the "portable kilns" of Neolithic Greece (Vitelli, this volume). These "kilns" fit the definition when they actually contain combustion; otherwise they seem more like early (and large) saggars.

In any case, I would now modify my earlier definition of "kiln" to incorporate these varying perspectives and principles, past and present, and to make it more "archaeology friendly." A revised definition of a ceramic kiln might better designate "enclosed or semi-enclosed chamber(s), usually relatively permanent or re-usable and non-portable, to hold ceramic materials during firing and to generate, contain, and channel heat for that purpose."

Attempting to outline the ancient path of evolution of these kiln technologies and firing strategies is an interesting exercise, but it is evident from the categorizations in my introductory chapter and from all the contributions in this volume that such an exercise is fraught with peril. There is little to no evidence for a simple evolutionary sequence of increasing efficiency and "progress" in pottery firing technology. Instead, what emerges from these papers are two intriguing points.

First, there is an extremely long temporal continuity and spatial distribution of some kiln types. Many authors noted continuities between the prehistoric kilns they excavated and kilns used by traditional potters living nearby today. Such long-standing kiln types were evident, for example, in highland and Gulf coastal Mexico from the Classic period up to modern times (Feinman and Balkansky, this volume; Pool, this volume). Even more striking is the temporal continuities in tile kilns, in which little change occurred from kilns in ancient Rome to those of its colony in Spain, and thence to the Spanish Colonial New World and up to the modern era (Costello, this volume).

Second, and most damaging to unilineal "progressivist" schemes of kiln evolution, several papers in this volume indicate that multiple types of firing arrangements were in more-or-less simultanous use in the same regions, or even at the same sites, in prehistory. Open firings, pit firings, and kiln firings often coexisted, and they seem to have been used for firing different products (see, e.g., contributions by Miller; Feinman and Balkansky; Shimada). Different firing strategies not only present different advantages depending on the wares being fired, they also vary with the presence of certain kinds or components of the kiln firing structure.

For example, relatively coarse, low-fired, unglazed and unvitrified ware (known as terracotta or earthenware) is generally characteristic of very early pottery throughout the world, and also describes nearly all New World pottery prior to European conquest. These wares may undergo some minor sintering but as they lack glazes there are few problems with vessels sticking together as a consequence of sintering or vitrification. Open firing systems with little or no formal structures work reasonably well for these wares. In addition, they undergo comparatively minor shrinkage, and are less likely to shift and damage each other during firing. For these kinds of wares, Kingery's design principle #4 (stable setting for the ware) may be less significant.

However, a stable setting for fragile finewares and, even more importantly, the physical separation of fuel and wares has some advantages. One is that such separation prevents movement of vessels as fuel is consumed during firing. Furthermore, while coarse wares generally can be fired in an open fire because they are able to sustain the rapid heating and thermal shocking, fine-paste wares benefit from slower heating and some barrier giving protection from direct contact with the flame. Gosselain (1992:246) notes that there may be "an inverse correlation between the heating rate and the presence of a material separating the pots from the fuel .... [and this] might explain, more than the temperatures reached or the thermic homogeneity, the choices and innovations in the field of firing techniques ...."

Stable settings and separation of fuel from wares are particularly important for glazed and vitrified ceramics. These wares need a longer soak at high temperatures and thus benefit not only from a stable platform but also a separate chamber to better contain the heat.

In any event, the specific order of advances in kiln development as set out by Rhodes seems too simplistic to accommodate the complexities increasingly being uncovered in the archaeological record. Nonetheless, there may be some reality to the general trend, in that it roughly approximates some gross trends of prehistoric pottery making: increasingly finer wares, higher temperatures, and ultimately vitrification.

MULTIPLE FIRING STRATEGIES

We need to return for a moment to the implications of multiple firing strategies for understanding pottery production in prehistory. Papers in this volume primarily treat manufacture of unglazed earthenwares (e.g., Vitelli; Miller; Blinman and Swink; Shimada; Feinman and Balkansky; Pool; Urban et al.; Costello), and also production at the household or "household industry" level.

This location and type of production organization is traditionally the most difficult to investigate archaeologically because (it is assumed that) there was little capital investment in physical facilities (such as kilns and workshops) in comparison to nucleated "industrial" production. A corollary assumption is that capital investment in permanent kiln structures is not warranted unless dictated by large volumes of production (e.g., for urban areas). For example, Feinman and Balkansky cite Sheehy (1992) in commenting on this relationship between volume of production and type of firing: low volume production uses open or pit strategies, while updraft and more complex firing technologies are used in those areas with greater demand.

But these assembled papers have clearly shown that specialized firing facilities--kilns--seem to have developed unexpectedly early in many areas world. As these papers amply demonstrate, pit kilns and other impermanent (and often partially or wholly subterranean) firing features were surprisingly widespread, especially in areas presumed to have carried out firings without kilns. (Possibly the common assumption of a very late development of kilns in pottery production refers specifically to the late development of complex permanent multi-chambered structures.) But it is evident now that simple assumptions, such as those underlying unilineal evolutionary schemes for ceramic pyrotechnologies, and direct correlations between volume/demand and technological complexity, are oversimplifications of complicated phenomena. Disentangling the various economic, cultural, and technological strands that comprise this web will take some time. For the moment I would only say that we need to spend more time incorporating organizational and socioeconomic (beyond demand) factors that play a role in pottery production. For example, it appears from some of these papers (e.g., Miller; Pool) that different kinds of firing strategies were used for different kinds of wares, based not only on volume of demand or intensity of production. Types of firings also varied on the basis of the wares' technological characteristics (coarse vs. fine pastes), for whom they were produced (elite/prestige use, domestic/utilitarian use; local vs. exported), and by whom they were produced ("attached" specialists, general household producers, "industrial" specialists). Some of these arrangements might very well involve inverse correlations between firing style and volume or intensity of production; they might also vary from community to community within a particular socioeconomic network, especially if there is community-based specialization in wares/forms.

IMPLICATIONS FOR ARCHAEOLOGICAL RESEARCH STRATEGIES

A final set of questions to evolve from all these considerations concerns how archaeologists might advance the study of early pottery firing. Open, bonfire, and pit-kiln firing techniques normally leave few surface traces for archaeologists to recover as a basis for reconstructing production strategies. What strategies can archaeologists employ to better recover this seemingly ephemeral evidence? What should archaeologists look for when above-ground kilns are not apparent, and pit-kiln firing might be suspected?

Pit kilns, particularly if they are small and/or only briefly used, and buried by overburden, are nearly impossible to recognize in surface survey. Similarly, even more formally constructed kilns, such as updraft kilns, may be semi-subterranean; often only the base is preserved because the ware chamber walls may have been damaged or intentionally destroyed after the accumulated thermal stresses of multiple use. On the other hand, double-chambered updraft kilns (and some single-chambered examples) may be more readily preserved because higher sustained firing temperatures make the upper chamber walls harder and more resistant to erosive damage.

In surface survey, common factors that are incorporated into inferences of pottery production at a particular site or in a particular area include (see Stark 1985; Rice 1987:177-180): location of nearby high-quality ceramic resources; location of nearby potting communities; spatial distribution of ceramic styles ("criterion of abundance"); occurrence of artifacts related to ceramic production (polishing; stones, molds, wheels, etc.); indications of burning - large areas of reddened soil, deposits of ash, thermally altered rock; occurrence of kiln furniture - stilts, props, saggars; occurrence of firing wasters (but see Urban et al., this volume); occurrence of unfired vessels of limited style or shape. Four papers in this volume suggest additional indicators that archaeologists might find on the surface, indicators that may be specific to pit kilns and simple updraft kilns, and both relate to temporary or ad hoc roofing applied to improve heat retention.

One of these indicators consists of distinctive fired adobe fragments that remain from temporary clay seals or "roofs" to enclose the top of the kilns to retain heat. Pool, for example, observed surface distributions of fiber-tempered adobe fragments of kiln construction at Matacapan, and Feinman and Balkansky noted burned "concretions" from the roof of a replicated kiln (see also Costello). Urban and her students were able to differentiate two different kinds, sizes, and shapes of *bajareque* (daub) fragments at La Sierra, depending on whether the daub was used in house or kiln construction. "Kiln bajareque" was porous, fire-tempered, and often vitrified; pieces were large, thick, and often curved. "House bajareque" was soft, sandy, easily eroded, and usually in small pieces.

Other remnants of firings in open-topped pit or updraft kiln are "cover sherds." As Blinman and Swink, Feinman and Balkansky, and Shimada note in their contributions, another way of creating temporary roofing for kilns is by topping the wares with sherds to help hold in the heat of firing. Concentrations of these large sherds, perhaps unusually brittle and/or partially vitrified from repeated subjection to high temperatures, are another item to add to the list of surface indicators of prehistoric firing locales.

Archaeologists seeking evidence of pottery production also need to rethink some of their general excavation strategies. In some cases, particularly in heavily settled and/or urban areas (see contributions by Miller; Pool; Shimada; Urban et al.; also Arnold 1991), kilns may be located in residential/workshop compounds. Even in instances of simpler household production, open or pit firings may take place in patios or courtyards of domestic units. In such situations, the possibility of identification of these firing features is reasonably good given relatively intensive excavations. However, when firings take place outside of household compounds, the probability of encountering these features seems largely a matter of chance unless there are clear surface indicators. In any of these cases, the likelihood of recovery of firing features can only be enhanced by revising excavation strategies to include broad exposures, as Feinman and Balkansky urge. If such features are difficult to find in Oaxaca, Mexico, which is arid and has relatively little vegetation cover, they will be even more elusive in heavily vegetated areas.

Finally, several papers (Shimada; Costello; McCray) call attention to the need for more integrated, multidisciplinary investigations into ancient firing kilns and firing strategies. Such investigations should incorporate documentary studies, archaeology, ethnoarchaeology, replicative experiments, and archaeometric analyses. Overall, such integrated approaches are to be heartily applauded, for they certainly yield broad insights into pottery manufacture and pyrotechnologies. However, they are not without some disadvantages. Perhaps the first is cost: the expenses of consultants and collaborators for their expertise and field and lab analysis fees may be nearly prohibitive unless some special arrangements can be made. In addition, archaeometric studies are perhaps best directed to questions other than firing temperature, given the variability that occurs both between and within different firing strategies (see Gosselain 1992). And some authors have cautioned against "the tyranny of the ethnographic record," noting that uncritical applications of ethnographic analogy can sometimes obscure some kinds of variability or confuse past and very recent behaviors (Feinman and Balkansky, this volume; also Rice 1996).

CONCLUDING REMARKS

All this having been said, this volume makes no pretense of covering all types of kilns in all times and places in prehistory and history. The papers focused on ceramic kilns wity one contribution on glass making and glass furnaces, and by editorial fiat ignored other types of kilns such as calcination kilns, and kilns for firing other types of materials. Even within the topic of ceramic kilns the coverage here is admittedly limited. Rather than being encylopedic in scope, this volume was intended to be a compendium of case studies for two audiences. One audience consists of archaeologists present and future who find themselves working with kilns: for them it is intended as a reference, with the additional hope that the assembled papers may serve as a stimulus to further research. The second audience is ceramic engineers, industry specialists, and modern potters interested in the history of technology: for them it is my hope that the volume will prompt a greater appreciation for the difficulties faced by prehistoric potters in firing their wares, and the variety of solutions they achieved.

For both audiences, it would be useful to develop additional volumes on kilns and firings, for example more systematic coverage of ethnoarchaeological studies, analysis of calcination kilns and plaster making (since the use of plaster was widespread throughout the world, past and present), and study of kilns and firings as explored by historical/industrial archaeologists. The admittedly limited coverage of the present volume also would be remedied by additional consideration of glassware, non-modern industrial brick/tile manufacture, and high-temperature glazed wares such as stonewares and porcelain.

To conclude, the papers in this volume show that prehistoric potters developed a variety of creative technical solutions to Kingery's basic design principles. In so doing they incorporated careful consideration of resources (clays, fuel), local weather and topography, wares to be fired, "market share," and so forth to try to assure predictable and reliable firings, and "better" products. We can only hope that future research will continue to illuminate the many questions that still remain about prehistoric and historic pottery kilns.

REFERENCES

Arnold, P.
1991 *Domestic Ceramic Production and Spatial Organization: A Mexican Case Study in Ethnoarchaeology.* Cambridge University Press, Cambridge.

Gosselain, O.P.
1992 Bonfire of the enquiries. Pottery firing temperatures in archaeology: What for? *Journal of Archaeological Science* 19(3):243-259.

Morgan, L.H.
1964[1877] *Ancient Society.* Belknap Press, Cambridge, MA.

Rhodes, D.
1981 *Kilns. Design, Construction, and Operation* (2nd ed.). Chilton Book Company, Radnor, PA.

Rice, P.M.
1987 *Pottery Analysis, A Sourcebook.* University of Chicago Press, Chicago, IL.
1996 Recent Ceramic Analysis: 2. Composition, Production, and Theory. *Journal of Archaeological Research* 4(3):165-202.

Rye, O.S., and C. Evans
1976 *Traditional Pottery Techniques of Pakistan: Field and Laboratory Studies.* Smithsonian Contributions to Anthropology no. 21. Smithsonian Institution, Washington, D.C.

Sheehy, J.J.
1992 *Ceramic Production in Ancient Teotihuacan, Mexico: A Case Study of Tlajinga 33.* Unpublished Ph.D. dissertation, The Pennsylvania State University, University Park.

Stark, B.L.
1985 Archaeological identification of pottery-production locations: Ethnoarchaeological data in Mesoamerica. In, *Decoding Prehistoric Ceramics*, B.A. Nelson, ed., pp. 158-194. Southern Illinois University Press, Carbondale.

KEYWORD AND AUTHOR INDEX

Printed and bound by CPI Group (UK) Ltd, Croydon, CR0 4YY

24/08/2026

14947831-0002